ANNUAL EDITIONS

Geography

Twenty-Third Edition

EDITOR

Gerald R. Pitzl

Macalester College (Professor Emeritus)

Gerald R. Pitzl received his bachelor's degree in secondary social science education from the University of Minnesota in 1964 and his MA (1971) and PhD (1974) in geography from the same institution. Dr. Pitzl has taught a wide array of geography courses, and he is the author of a number of articles on geography, the developing world, and the use of the Harvard Case Method. His book, *Encyclopedia of Human Geography,* was published in January 2004. This one-volume work is designed for use in the Advanced Placement (AP) Program in human geography and other courses in the discipline. Dr. Pitzl continues to conduct workshops and in-service sessions on active learning and case-based discussion methods. Since 2002, he has served as an educational consultant with the New Mexico Public Education Department in Santa Fe. His most recent assignment is with the Rural Education Division.

Higher Education

Boston   Burr Ridge, IL   Dubuque, IA   New York   San Francisco   St. Louis
Bangkok   Bogotá   Caracas   Kuala Lumpur   Lisbon   London   Madrid   Mexico City
Milan   Montreal   New Delhi   Santiago   Seoul   Singapore   Sydney   Taipei   Toronto

ANNUAL EDITIONS: GEOGRAPHY, TWENTY-THIRD EDITION

1 2 3 4 5 6 7 8 9 0 QWD/QWD 0 9

ISBN 978–0–07–351551–9
MHID 0–07–351551–5
ISSN 1091–9937

Managing Editor: *Larry Loeppke*
Senior Managing Editor: Faye Schilling
Developmental Editor: *Debra A. Henricks*
Editorial Coordinator: *Mary Foust*
Editorial Assistant: *Cindy Hedley*
Production Service Assistant: *Rita Hingtgen*
Permissions Coordinator: *Lenny J. Behnke*
Senior Marketing Manager: *Julie Keck*
Marketing Communications Specialist: *Mary Klein*
Marketing Coordinator: *Alice Link*
Project Manager: *Joyce Watters*
Design Specialist: *Tara McDermott*
Cover Graphics: *Kristine Jubeck*

Compositor: Laserwords Private Limited
Cover Image: © Getty Images/RF (inset); © Comstock Images/Alamy/RF (background)

Library in Congress Cataloging-in-Publication Data
Main entry under title: Annual Editions: Geography. 2010/2011.
  1. Geography—Periodicals. I. Pitzl, Gerald R., *comp*. II. Title: Geography.
658'.05

www.mhhe.com

Editors/Academic Advisory Board

Members of the Academic Advisory Board are instrumental in the final selection of articles for each edition of ANNUAL EDITIONS. Their review of articles for content, level, and appropriateness provides critical direction to the editors and staff. We think that you will find their careful consideration well reflected in this volume.

Preface

In publishing ANNUAL EDITIONS we recognize the enormous role played by the magazines, newspapers, and journals of the public press in providing current, first-rate educational information in a broad spectrum of interest areas. Many of these articles are appropriate for students, researchers, and professionals seeking accurate, current material to help bridge the gap between principles and theories and the real world. These articles, however, become more useful for study when those of lasting value are carefully collected, organized, indexed, and reproduced in a low-cost format, which provides easy and permanent access when the material is needed. That is the role played by ANNUAL EDITIONS.

The articles in this twenty-third edition of *Annual Editions: Geography* represent the wide range of topics associated with the discipline of geography. The major themes of spatial relationships, regional development, population dynamics, and socioeconomic inequalities exemplify the diversity of research areas within geography.

The book is organized into five units, each containing articles relating to geographical themes. Selections address the conceptual nature of geography and the global and regional problems in the world today. This latter theme reflects the geographer's concern with finding solutions to these serious issues. Regional problems, such as food shortages in the Sahel and global warming, concern not only geographers but researchers from a wide array of other disciplines.

The association of geography with other fields is important because expertise from related research will be necessary in finding solutions to difficult global and regional problems. Input from the geography's conceptual base is vital to our common search for solutions. The discipline has always been integrative. That is, geography uses evidence from many sources to answer the basic questions, "Where is it?" "Why is it there?" and "What is its relevance?" The first group of articles emphasizes the interconnectedness not only of places and regions in the world but of efforts toward solutions to problems as well. No single discipline can have all of the answers to the problems facing the world today; the complexity of the issues is simply too great.

The writings in Unit 1 discuss particular aspects of geography as a discipline and provide examples of the topics presented in the remaining four sections. Units 2, 3, and 4 represent major themes in geography. Unit 5 addresses important global and regional problems dealt with by geographers and other investigators.

Annual Editions: Geography, 23rd *edition,* will be useful to both faculty members and students in their study of geography. The anthology is designed to provide detail and case study material to supplement the standard textbook treatment of geography. The goals of this anthology are to introduce students to the richness and diversity of topics relating to places and regions on the Earth's surface, to pay heed to the serious problems facing humankind, and to stimulate the search for more information on topics of interest. As such, this anthology is an ideal companion volume for use both in introductory college courses and the **Advanced Placement (AP) Program in Human Geography** found in high schools across the country. This program, like others sponsored by The College Board, has grown steadily since its inception. Currently, over 18,000 high school students are enrolled in the AP Human Geography Program.

I would like to express my gratitude to Devi Benjamin for her encouragement and invaluable assistance in preparing the twenty-third edition. Without her enthusiasm and professional efforts, this project would not have moved along as efficiently as it did. Special thanks are also extended to the McGraw-Hill/Dushkin editorial staff for coordinating the production of the reader. A word of thanks must go as well to all those who recommended articles for inclusion in this volume and who commented on its overall organization.

In order to improve the next edition of *Annual Editions: Geography,* we need your help. Please share your opinions by filling out and returning to us the postage-paid **ARTICLE RATING FORM** found on the last page of this book. We will give serious consideration to all your comments and recommendations.

Gerald R. Pitzl
Editor

Contents

UNIT 1
Geography in a Changing World

The concepts in bold italics are developed in the article. For further expansion, please refer to the Topic Guide.

UNIT 2
Human-Environment Relations

The concepts in bold italics are developed in the article. For further expansion, please refer to the Topic Guide.

UNIT 3
The Region

The concepts in bold italics are developed in the article. For further expansion, please refer to the Topic Guide.

UNIT 4
Spatial Interaction and Mapping

The concepts in bold italics are developed in the article. For further expansion, please refer to the Topic Guide.

UNIT 5
Population, Resources, and Socioeconomic Development

The concepts in bold italics are developed in the article. For further expansion, please refer to the Topic Guide.

The concepts in bold italics are developed in the article. For further expansion, please refer to the Topic Guide.

Correlation Guide

The *Annual Editions* series provides students with convenient, inexpensive access to current, carefully selected articles from the public press. **Annual Editions: Geography 23/e** is an easy-to-use reader that presents articles on important topics such as *human-environment relations, regional geography, population,* and many more. For more information on *Annual Editions* and other *McGraw-Hill Contemporary Learning Series* titles, visit www.mhcls.com.

This convenient guide matches the units in **Annual Editions: Geography 23/e** with the corresponding chapters in two of our best-selling McGraw-Hill Geography textbooks by Fellmann et al. and Bradshaw et al.

Annual Editions: Geography 23/e	**Human Geography: Landscapes of Human Activities, 11/e by Fellmann et al.**	**Contemporary World Regional Geography, 3/e by Bradshaw et al.**
Unit 1: Geography in a Changing World	**Chapter 10:** Patterns of Development and Change	**Chapter 1:** Globalization and World Regions **Chapter 13:** A World of Geography
Unit 2: Human-Environment Relations	**Chapter 4:** Population **Chapter 13:** Human Impacts on Natural Systems	**Chapter 2:** Concepts in World Regional Geography **Chapter 4:** Russia and Neighboring Countries **Chapter 9:** Africa South of the Sahara **Chapter 10:** Australia, Oceania, and Antarctica
Unit 3: The Region	**Chapter 1:** Introduction **Chapter 2:** Roots and Meaning of Culture	**Chapter 1:** Globalization and World Regions **Chapter 2:** Concepts in World Regional Geography
Unit 4: Spatial Interaction and Mapping	**Chapter 1:** Introduction **Chapter 3:** Spatial Interaction and Spatial Behavior **Chapter 12:** The Political Ordering of Space	**Chapter 1:** Globalization and World Regions **Chapter 6:** Southeast Asia **Chapter 8:** Northern Africa and Southwest Asia
Unit 5: Population, Resources, and Socioeconomic Development	**Chapter 4:** Population **Chapter 8:** Livelihood and Economy: Primary Activities **Chapter 9:** Livelihood and Economy: From Blue Collar to Gold Collar **Chapter 10:** Patterns of Development and Change	**Chapter 3:** Europe **Chapter 4:** Russia and Neighboring Countries **Chapter 6:** Southeast Asia **Chapter 8:** Northern Africa Southwestern Asia **Chapter 9:** Africa South of the Sahara **Chapter 12:** North America

Topic Guide

This topic guide suggests how the selections in this book relate to the subjects covered in your course. You may want to use the topics listed on these pages to search the Web more easily.

On the following pages a number of websites have been gathered specifically for this book. They are arranged to reflect the units of this Annual Editions reader. You can link to these sites by going to *http://www.mhcls.com*.

All the articles that relate to each topic are listed below the bold-faced term.

Internet References

The following Internet sites have been selected to support the articles found in this reader. These sites were available at the time of publication. However, because websites often change their structure and content, the information listed may no longer be available. We invite you to visit *http://www.mhcls.com* for easy access to these sites.

Annual Editions: Geography, 23/e

General Sources

About: Geography
http://geography.about.com

This website, created by the About network, contains hyperlinks to many specific areas of geography, including cartography, population, country facts, historic maps, physical geography, topographic maps, and many others.

The Association of American Geographers (AAG)
http://www.aag.org

Surf this site of the Association of American Geographers to learn about AAG projects and publications, careers in geography, and information about related organizations.

Geography Network
http://www.geographynetwork.com

The Geography Network is an online resource to discover and access geographical content, including live maps and data, from many of the world's leading providers.

International Geographic Union
http://www.igu-net.org/uk/igu.html

Website of the International Geographic

National Geographic Society
http://www.nationalgeographic.com

This site provides links to National Geographic's huge archive of maps, articles, and other documents. Search the site for information about worldwide expeditions of interest to geographers.

The New York Times
http://www.nytimes.com

Browsing through the archives of the *New York Times* will provide you with a wide array of articles and information related to the different subfields of geography.

Social Science Internet Resources
http://library.wcsu.edu/web/

This site is a definitive source for geography-related links to universities, browsers, cartography, associations, and discussion groups.

U.S. Geological Survey (USGS)
http://www.usgs.gov

This site and its many links are replete with information and resources for geographers, from explanations of El Niño, to mapping, to geography education, to water resources. No geographer's resource list would be complete without frequent mention of the USGS.

UNIT 1: Geography in a Changing World

Alternative Energy Institute (AEI)
http://www.altenergy.org

The AEI will continue to monitor the transition from today's energy forms to the future in a "surprising journey of twists and turns." This site is the beginning of an incredible journey.

The Science Mission Directorate Website: Mission to Planet Earth
http://science.hq.nasa.gov/

This site will direct you to information about NASA's Mission to Planet Earth program and its Science of the Earth System. Surf here to learn about satellites, El Niño, and even "strategic visions" of interest to geographers.

Poverty Mapping
http://www.povertymap.net

Poverty maps can quickly provide information on the spatial distribution of poverty. Here you will find maps, graphics, data, publications, news, and links that provide the public with poverty mapping from the global to the subnational level.

UNIT 2: Human-Environment Relations

Alliance for Global Sustainability (AGS)
http://www.global-sustainability.org

The AGS is a cooperative venture seeking solutions to today's urgent and complex environmental problems. Research teams from four universities study large-scale, multidisciplinary environmental problems that are faced by the world's ecosystems, economies, and societies.

California Climate Change Portal
http://www.climatechange.ca.gov

Tracks aspects of climate change.

Human Geography
http://www.le.ac.uk/gg/cti/

The CTI Centre for Geography, Geology, and Meteorology provides this site, which contains links to human geography in relation to agriculture, anthropology, archaeology, development geography, economic geography, geography of gender, and many others.

The North-South Institute
http://www.nsi-ins.ca

Searching this site of the North-South Institute—which works to strengthen international development cooperation and enhance gender and social equity—will help you find information on a variety of development issues.

United Nations Environment Programme (UNEP)
http://www.unep.ch

Consult this home page of UNEP for links to critical topics of concern to geographers, including desertification and the impact of trade on the environment. The site will direct you to useful databases and global resource information.

U.S. Global Change Research Program
http://www.usgcrp.gov

This government program supports research on the interactions of natural and human-induced changes in the global environment and their implications for study. Find details on the atmosphere, climate change, global carbon and water cycles, ecosystems, and land use plus human contributions and responses.

World Health Organization
http://www.who.int

This home page of the World Health Organization will provide you with links to a wealth of statistical and analytical information about health in the developing world.

Internet References

UNIT 3: The Region

AS at UVA Yellow Pages: Regional Studies
http://xroads.virginia.edu/~YP/regional.html

Those interested in American regional studies will find this site a gold mine. Links to periodicals and other informational resources about the Midwest/Central, Northeast, South, and West regions are provided here.

Can Cities Save the Future?
http://www.huduser.org/publications/econdev/habitat/prep2.html

This press release about the second session of the Preparatory Committee for Habitat II is an excellent discussion of the question of global urbanization.

Minnesota Department of Transportation
www.dot.state.mn.us

Photos of the interstate bridge re-opening with webcam coverage and photographs of workers.

NewsPage
http://www.individual.com

Individual, Inc., maintains this business-oriented website. Geographers will find links to much valuable information about such fields as energy, environmental services, media and communications, and health care.

World Regions & Nation States
http://www.worldcapitalforum.com/worregstat.html

This site provides strategic and competitive intelligence on regions and individual states, geopolitical analyses, geopolitical factors of globalization, geopolitics of production, and much more.

UNIT 4: Spatial Interactions and Mapping

Edinburgh Geographical Information Systems
http://www.geo.ed.ac.uk/home/gishome.html

This valuable site, hosted by the Department of Geography at the University of Edinburgh, provides information on all aspects of Geographic Information Systems and provides links to other servers worldwide. A GIS reference database as well as a major GIS bibliography is included.

Geography for GIS
http://www.ncgia.ucsb.edu/cctp/units/geog_for_GIS/GC_index.html

This hyperlinked table of contents was created by Robert Slobodian of Malaspina University. Here you will find information regarding GIS technology.

GIS Frequently Asked Questions and General Information
http://factfinder.census.gov/home/saff/main.html

Browse through this site to get answers to FAQs about Geographic Information Systems. It can direct you to general information about GIS as well as guidelines on such specific questions as how to order U.S. Geological Survey maps. Other sources of information are also noted.

International Map Trade Association
http://www.maptrade.org

The International Map Trade Association offers this site for those interested in information on maps, geography, and mapping technology. Lists of map retailers and publishers as well as upcoming IMTA conferences and trade shows are noted.

PSC Publications
http://www.psc.isr.umich.edu

Use this site and its links from the Population Studies Center of the University of Michigan for spatial patterns of immigration and discussion of white and black flight from high immigration metropolitan areas in the United States.

UNIT 5: Population, Resources, and Socioeconomic Development

African Studies WWW U.Penn
http://www.sas.upenn.edu/African_Studies/AS.html

Access to rich and varied resources that cover such topics as demographics, migration, family planning, and health and nutrition is available at this site.

Capitol Region Council of Governments (CRCOG)
www.crcog.org/gissearch

Access detailed GIS map coverage for municipalities in Connecticut.

Geography and Socioeconomic Development
http://www.ksg.harvard.edu/cid/andes/Documents/Background %20Papers/Geography&Socioeconomic%20Development.pdf

John L. Gallup wrote this 19-page background paper examining the state of the Andean region. He explains the strong and pervasive effects geography has on economic and social development.

Human Rights and Humanitarian Assistance
http://www.etown.edu/vl/humrts.html

Through this site, part of the World Wide Web Virtual Library, you can conduct research into a number of human-rights topics in order to gain a greater understanding of the issues affecting indigenous peoples in the modern era.

Hypertext and Ethnography
http://www.umanitoba.ca/faculties/arts/anthropology/tutor/aaa_ presentation.new.html

This site, presented by Brian Schwimmer of the University of Manitoba, will be of great value to people who are interested in culture and communication. He addresses such topics as multivocality and complex symbolization, among many others.

Research and Reference Library of Congress
http://lcweb.loc.gov/rr/

This research and reference site of the Library of Congress will lead you to invaluable information on different countries. It provides links to numerous publications, bibliographies, and guides in area studies that can be of great help to geographers.

Space Research Institute
http://arc.iki.rssi.ru/eng/

Browse through this home page of Russia's Space Research Institute for information on its Environment Monitoring Information Systems, the IKI Satellite Situation Center, and its Data Archive.

Worldmapper
www.worldmapper.org

Presents regional maps.

World Population and Demographic Data
http://geography.about.com/cs/worldpopulation/

On this site you will find information about world population and additional demographic data for all the countries of the world.

Telegraph.co.uk
http://www.telegraph.co.uk/travel/picturegalleries/

Examples of cartograms from the book *The Atlas of the Real World*.

World Map

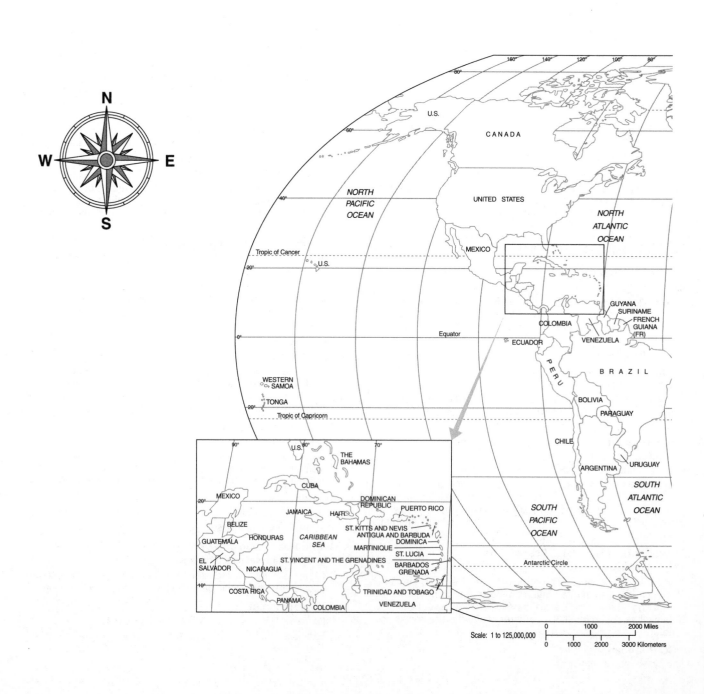

N
W   E
S

160° 140° 120° 100° 80°
80°

U.S.

CANADA

60°

NORTH
PACIFIC
OCEAN

40°

UNITED STATES

NORTH
ATLANTIC
OCEAN

Tropic of Cancer

MEXICO

20°
U.S.

GUYANA
SURINAME
FRENCH
GUIANA
(FR)

COLOMBIA

VENEZUELA

Equator

ECUADOR

0°

BRAZIL

PERU

WESTERN
SAMOA

BOLIVIA

TONGA

PARAGUAY

20°
Tropic of Capricorn

CHILE

SOUTH
ATLANTIC
OCEAN

ARGENTINA

URUGUAY

SOUTH
PACIFIC
OCEAN

90° 80° 70°

U.S.

THE
BAHAMAS

CUBA

MEXICO

20°

DOMINICAN
REPUBLIC

PUERTO RICO

JAMAICA

HAITI

BELIZE

ST. KITTS AND NEVIS
ANTIGUA AND BARBUDA

GUATEMALA

HONDURAS

CARIBBEAN
SEA

DOMINICA

MARTINIQUE

ST. LUCIA

EL
SALVADOR

NICARAGUA

ST. VINCENT AND THE GRENADINES

BARBADOS
GRENADA

Antarctic Circle

10°

COSTA RICA

PANAMA

COLOMBIA

TRINIDAD AND TOBAGO

VENEZUELA

Scale: 1 to 125,000,000

0   1000   2000 Miles
0 1000 2000 3000 Kilometers

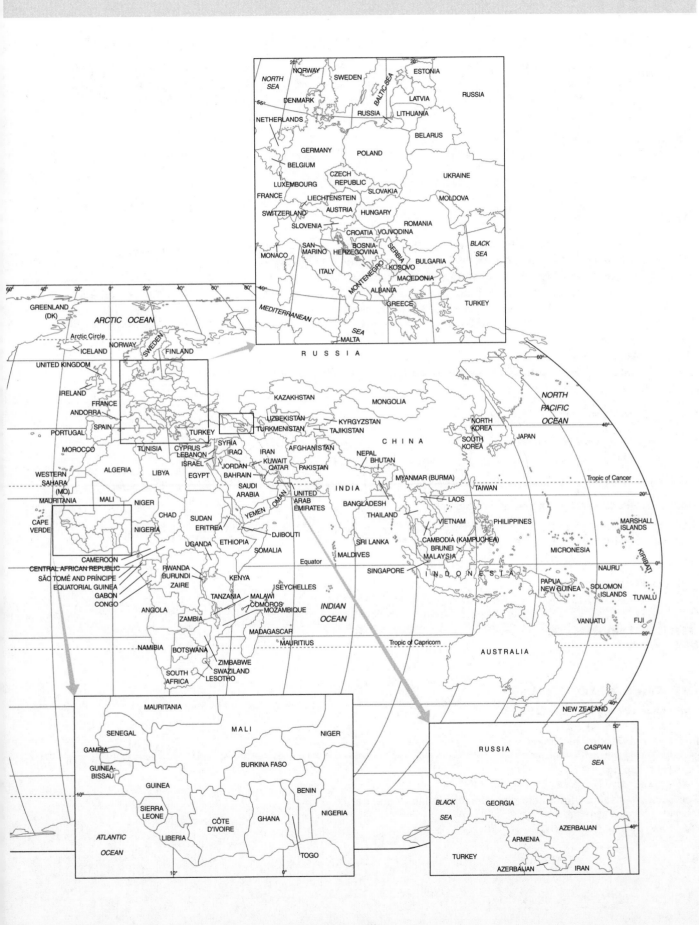

UNIT 1

Geography in a Changing World

Unit Selections

1. **The Big Questions in Geography,** Susan L. Cutter, Reginald Golledge, and William L. Graf
2. **Rediscovering the Importance of Geography,** Alexander B. Murphy
3. **The Four Traditions of Geography,** William D. Pattison
4. **The Changing Landscape of Fear,** Susan L. Cutter, Douglas B. Richardson, and Thomas J. Wilbanks
5. **The Geography of Ecosystem Services,** James Boyd
6. **The Agricultural Impact of Global Climate Change: How Can Developing-Country Farmers Cope?,** Nathan Russell
7. **When Diversity Vanishes,** Don Monroe
8. **Classic Map Revisited: The Growth of Megalopolis,** Richard Morrill

Key Points to Consider

- Why is geography called an integrating discipline?

- How is geography related to earth science? Give some examples of these relationships.

- What are area studies? Why is the spatial concept so important in geography? What is your definition of geography?

- How has GIS changed geography and cartography?

- What does interconnectedness mean in terms of places? Give examples of how you as an individual interact with people in other places. How are you "connected" to the rest of the world?

- How would you describe your personal "sense of place"?

- Why is diversity important?

Student Website
www.mhcls.com

Internet References

Alternative Energy Institute (AEI)
http://www.altenergy.org
International Geographic Union
http://www.igu-net.org/uk/igu.html
Poverty Mapping
http://www.povertymap.net
The Science Mission Directorate Website: Mission to Planet Earth
http://science.hq.nasa.gov/

What is geography? This question has been asked innumerable times, but it has not elicited a universally accepted answer, even from those who are considered to be members of the geography profession. The reason lies in the very nature of geography as it has evolved through time. Geography is an extremely wide-ranging discipline, one that examines appropriate sets of events or circumstances occurring at specific places. Its goal is to answer certain basic questions.

The first question—Where is it?—establishes the location of the subject under investigation. The concept of location is very important in geography, and its meaning extends beyond the common notion of a specific address or the determination of the latitude and longitude of a place. Geographers are more concerned with the relative location of a place and how that place interacts with other places both far and near. Spatial interaction and the determination of the connections between places are important themes in geography.

Once a place is "located," in the geographer's sense of the word, the next question is, Why is it here? For example, why are people concentrated in high numbers on the North China plain, in the Ganges River Valley in India, and along the eastern seaboard in the United States? Conversely, why are there so few people in the Amazon basin and the Central Siberian lowlands? Generally, the geographer wants to find out why particular distribution patterns occur and why these patterns change over time.

The element of time is another extremely important ingredient in the geographical mix. Geography is most concerned with the activities of human beings, and human beings bring about change. As changes occur, new adjustments and modifications are made in the distribution patterns previously established. Patterns change, for instance, as new technology brings about new forms of communication and transportation and as once-desirable locations decline in favor of new ones. For example, people migrate from once-productive regions such as the Sahel when a disaster such as drought visits the land. Geography, then, is greatly concerned with discovering the underlying processes that can explain the transformation of distribution patterns and interaction forms over time. Geography itself is dynamic, adjusting as a discipline to handle new situations in a changing world.

Geography is truly an integrating discipline. The geographer assembles evidence from many sources in order to explain a particular pattern or ongoing process of change. Some of this evidence may even be in the form of concepts or theories borrowed from other disciplines. The first article in this unit discusses what the authors consider to be the big questions in geography. The second considers geography's renaissance in U.S. education and its "rediscovery."

Throughout its history, four main themes have been the focus of research work in geography. These themes or traditions, according to William Pattison in "The Four Traditions of

© C. Borland/PhotoLink/Getty Images

Geography," link geography with earth science, establish it as a field that studies land/human relationships, engage it in area studies, and give it a spatial focus. Although Pattison's article appeared over forty years ago, it is still referred to and cited frequently today. Much of the geographical research and analysis engaged in currently would fall within one or more of Pattison's traditions. Nonetheless, new areas are also opening up for geographers.

The next selection was chosen from the highly acclaimed book, *The Geographical Dimensions of Terrorism,* The introductory chapter, "The Changing Landscape of Fear," states that geography can play a key role in both understanding and combating terrorism. The importance of ecosystems study is discussed in the next article.

A discussion of global climate change and its impact on agriculture follows. "When Diversity Vanishes" raises a caution flag on the potential loss of complex systems from ecologies to economies. Richard Morrill brings the famous map of Megalopolis up to date in the next article.

The Big Questions in Geography

In noting his fondness for geography, John Noble Wilford, science correspondent for *The New York Times,* nevertheless challenged the discipline to articulate those big questions in our field, ones that would generate public interest, media attention, and the respect of policymakers. This article presents our collective judgments on those significant issues that warrant disciplinary research. We phrase these as a series of ten questions in the hopes of stimulating a dialogue and collective research agenda for the future and the next generation of geographic professionals.

SUSAN L. CUTTER, REGINALD GOLLEDGE, AND WILLIAM L. GRAF

Introduction

At the 2001 national meeting of the Association of American Geographers (AAG) in New York City, the opening session featured an address by John Noble Wilford, science correspondent for *The New York Times.* In very candid language, Wilford challenged the discipline to articulate the big questions in our field—questions that would capture the attention of the public, the media, and policymakers (Abler 2001). The major questions posed by Wilford's remarks include the following: Are geographers missing big questions in their research? Why is the research by geographers on big issues not being reported? And what role can the AAG play in improving geographic contributions to address big issues?

First, geographers are doing research on some major issues facing modern society, but not all of them. Geographic thinking is a primary component of the investigation of global warming, for example. Products of that research seen by decision makers and the public often take the form of maps and remote sensing images that explain the geographic outcomes of climate change. Geographic approaches are at the heart of much of the analysis addressing natural and technological hazards, with public interaction taking place through the mapping media. Earthquake, volcanic, coastal, and riverine hazards are all subject to spatial analysis that has become familiar to the public. The terrorist attacks of 11 September 2001 have stimulated new interest in geographic information systems that can be used in response to hazardous events and as guidance in emergency preparedness and response (Figure 1).

In addition to these recent challenges, however, there are major issues that geographers are not addressing adequately at the present time, as illustrated by the accounting that follows in this article. A primary reason for the disconnect between capability to help solve problems and the application of those skills for many major issues is the sociology of the discipline of geography. The majority of AAG members, for example, are academicians, and their agendas and reward structures are targeted at specialized research deeply buried in paradigms that are obscure to decision makers and the public. Additionally, this social structure tends to lead geographic researchers into investigations on small problems that can be solved quickly, produce professional publications, and support a drive for promotion and tenure, rather than investigating more complex, bigger problems that are not easily or quickly solved and do not necessarily lead to academic publications of a type the genre usually demands.

With few exceptions, those geographers outside the university setting are scattered and work individually, in small groups, or as members of larger interdisciplinary teams for governmental agencies, businesses, or private organizations. Because there are few true "institutes" of geographic research, it is difficult to focus geographic energy on big problems. Many geographers in these settings are responding to immediate and short-term demands on their time and talents, rather than leading the larger-scale investigations.

The work in which many geographers engage to address major problems is not reported for two reasons: it does not fit the classic mold for the research journals where geographers get their greatest career awards, and work related to policy often emerges without attribution to the researchers of origin. A significant example that illustrates this point is the work of the Committee on Geography of the Board on Earth Sciences and Resources in the National Research Council (NRC). The committee oversees study committees, which produce geographic studies and reports to guide the federal government in a wide variety of issues that qualify as big questions. Recent work, contributed to primarily by geographers and accomplished from a geographic perspective, includes advice to the U.S. Geological Survey on reformulating its research programs to address geographic issues entangled in urban expansion, hazards, and mapping. In other cases, one study committee is producing direction

Figure 1   Manhattan, New York, before the terrorist attacks of 11 September 2001 (left), and after (right). Photos by S. L. Cutter.

for the federal government on what decision makers and the general public need to know about the world of Islam, while another is investigating transportation issues related to urban congestion and the development of livability indicators. Other geographers participate in the Water Science and Technology Board of the NRC, with recent contributions including the use of the watershed concepts in ecosystem management and the role of dams in the security of public water supplies. Another example involves a geographer-led multidisciplinary group to investigate spatial thinking, and another geographer led a major effort in global mapping. In all of these cases, geographers play a central role, but the product of their work is ascribed only to an organization (the NRC), and individuals are recognized only in lists of contributors. If the reports successfully influence policy, the decision makers who actuate that policy take credit for the process, rather than the original investigators who made the recommendations.

The AAG plays a role in stimulating the research that addresses big questions of importance to modern society by recognizing such work and publicizing it. It may be that individual researchers will be more willing to undertake such research if their work is recognized by their colleagues in the discipline as being important and worthy of praise. The AAG can influence the National Science Foundation, the National Endowment for the Humanities, the National Institutes of Health, the National Geographic Society, and other funding sources to channel attention and resources to individuals or teams examining the big questions. Individual geographers are not likely to be able to exert much influence, except when they serve on review panels for these organizations, but the AAG can exert its influence from its steady and visible presence in Washington.

In trying to identify those issues that might qualify as big questions (Table 1), we have included wide-ranging concepts that encompass some conceptual issues (such as scale), but also point out specific topical areas that seem to demand particular attention at the moment. We argue that such diverse big ideas belong in this accounting because, in the end, they are related to each other and mutually supportive. Some of these big questions may be obscure to the public, but most of them are familiar to researchers and policymakers alike, who have already begun to address them. There is little hope that any collection of big questions can identify problems of equal "bigness," but the ones

Table 1   Big Questions in Geography

1. What makes places and landscapes different from one another, and why is this important?
2. Is there a deeply held human need to organize space by creating arbitrary borders, boundaries, and districts?
3. How do we delineate space?
4. Why do people, resources, and ideas move?
5. How has the earth been transformed by human action?
6. What role will virtual systems play in learning about the world?
7. How do we measure the unmeasurable?
8. What role has geographical skill played in the evolution of human civilization, and what role can it play in predicting the future?
9. How and why do sustainability and vulnerability change from place to place and over time?
10. What is the nature of spatial thinking, reasoning, and abilities?

we have identified all seem to warrant teams of researchers and significant funding rather than following the discipline's usual mode of a single or small group of investigators with funding limited to one or two years in duration. The communication of geographic research findings to the public in thoughtful, useful ways represents a major challenge. This challenge, by itself, can also be regarded as yet another big problem facing the discipline. With these introductory comments in mind, we now turn to those questions that we feel are important for the geographic community to address.

What Makes Places and Landscapes Different from One Another and Why Is This Important?

This first question goes to the core of the discipline the relevance of similarities and differences among people, places, and regions. What is the nature of uneven economic development and what can geography contribute to understanding this phenomenon? More specifically, how can national and global policies be implemented in a world that is increasingly fragmented politically, socially, culturally, and environmentally?

To elaborate on this question, we accept an assumption that the human mind is not constructed to handle large-scale continuous chaos. Nor does it function optimally when dealing with large-scale perfect uniformity. Between these two extremes there is variability, which is the dominant characteristic of both the natural world and the human world. To understand the nature of physical and human existence, we need to examine the occurrence and distribution of variability in various domains. For geographers, this examination has involved exploring the nature of spatial distributions, patterns, and associations, examining the effects of scale, and developing modes of representation that best communicate the outcomes of these explorations. In the course of this search for understanding of the essentials of spatial variation, geographers have attempted to comprehend the interaction between physical and human environments, how people adapt to different environments, and how knowledge about human-environment relations can be communicated through appropriate representational media.

Even in the absence of humans, the earth and the phenomena found on this planet are incredibly diverse. Variability is widespread; uniformity is geographically restricted. Determining the nature and occurrence of variability and uniformity are at the heart of the discipline of geography. No other area of inquiry has, as its primary goal, discovering, representing, and explaining the nature of spatial variability in natural and human environments at scales beyond the microscopic and the figural (body space) such as vista, environmental, or gigantic and beyond (Montello 1993). Most geography has been focused on vista, environmental or gigantic scales, but some (e.g., cognitive behavioral) emphasizes figural scale. Finding patterns or trends towards regularity at some definable scale amidst this variability provides the means for generalizing, modeling, and transferring knowledge from one spatial domain to another.

Law-like and theoretical statements can be made, and confidence in the relevance of decisions and policies designed to cope with existence can be determined.

Among other things, geographers have repeatedly found, at some scales, spatial regularity in distributions of occurrences that seem random or indeed chaotic at other scales. Sometimes this results from selecting an appropriate scale and format for summarizing and representing information. Examples include using very detailed environmental-scale data to discover the topologic properties of stream networks, or establishing the regular and random components of human settlement patterns in different environments.

Realizing the spatial variability in all phenomena is a part of the naïve understanding of the world. Being able to explain the nature of variability is the academic challenge that drives the discipline of geography. Like other scientists, geographers examine variability in their search for knowledge and understanding of the world we live in, particularly in the human environment relations and interactions that are a necessity for our continued existence.

Is There a Deeply Held Human Need to Organize Space by Creating Arbitrary Borders, Boundaries, and Districts?

Humans, by their very nature, are territorial. As human civilizations grew from hunter gatherers to more sedentary occupations, physical manifestation of the demarcation of space ensued. Hadrian's Wall kept the Scots and Picts out, the Great Wall of China protected the Ming Dynasties from the Mongols, and the early walled cities of Europe protected those places from barbarians and other acquisitive sociocultural groups.

At a more limited scale, internal spaces in cities were also divided, often based on occupation and/or class. As civilizations grew, space was organized and reorganized into districts that supported certain economic activities. City-states begat nation-states, and eventually most of the world was carved up into political spaces. Nation-states required borders and boundaries (all involving geography), as land and the oceans (and the resources contained within) were carved up into non-equal units. Within nations, land partitioning has been a factor in the decline of environmental quality. For example, the erection of barbed-wire fencing on the Great Plains to separate farming and ranching homesteads from each other did more to hasten the decline of indigenous species and landscape degradation than any other invention at the time (Worster 1979, 1993).

The modern equivalent of the human need and desire for delineating space is the notion of private property. Suburban homes with tall fences between neighbors, for example, help foster the ideal of separation from neighbors and disengagement from the community, both predicated on the need to protect "what's mine" (and of course the ubiquitous property value) as well as providing a basic need for privacy. The tendency for the rich to get richer and the poor to get poorer also applies to the values of these divided properties. The diffusion of democratically controlled, market based economies to much of the globe increases the significance of research that explores why we divide space. Pressing research questions include, for example: Are ghettos bursting with poverty-level inhabitants an inevitable consequence of democratic capitalist societies? Are such societies amenable to concerns for social justice? And how would such concerns influence the patterns and distributions of living activities?

We also lack some of the basic understanding of how the physical delineation of space affects our perception of it. Furthermore, we need better knowledge of how perceptions of physical space alter social, physical, and environmental processes. Finally, has globalization changed our view of the social construction of space? Does physical space still support spatial relations and spatial interactions, or are they becoming somewhat independent, as may be the case in social space, intellectual space, and cyberspace? How will the interactions between people, places, and regions change as our view of space (and time, for that matter) changes?

How Do We Delineate Space?

Once we understand *why* we partition space, we face a closely related issue: *how* do we do it? The definition of regions by drawing boundaries is deceptively simple. The criteria by which we delineate space have far-reaching consequences, because the resulting divisions of space play a large role in determining how we perceive the world. A map of the United States showing the borders of the states, for example, evokes a very different perception of the nation than a similar scale map showing the borders of the major river basins. A further difference in perception is created if the map shows major rivers as networks rather than as basins, and the resulting difference between perception of networks and perception of regions can direct knowledge and its application in divergent ways (National Research Council 1999, x). For example, should we conduct pollution oriented research on rivers or on watersheds, or on the state administrative units that potentially might control pollution? What are the implications of our choice of geographic framework?

The logical, rational delineation of spaces on the globe depends on the criteria to be used, but geographic research offers few established, widely accepted rules about what these criteria should be or how they might be employed. The designation of political boundaries without respect to ethnic cultures has wrought havoc in much of post colonial Africa and central Europe, for example, but geographers have not yet offered workable alternatives that account for the complexities of multicultural populations. In natural-science research and management, a major issue is the establishment of meaningful regions that can be aggregated together to scale up, or that can be disaggregated to scale down. Natural scientists also experience significant difficulty in designing compatible regions across topical subjects. For example, the blending of watersheds, ecosystems, and ranges of particular species poses significant problems in environmental management. Adding to the complexity from a management and policy perspective is the tangle of administrative regions, whose boundaries are often derived from political boundaries

rather than natural ones. Recognition of these problems is easy but offering thoughtful geographic solutions to them is not.

Geographers have much to contribute to the delineation of space by developing new knowledge and techniques for defining subdivisions of earth space based on specific criteria, including economic efficiency, compatibility across applications, ease of aggregation and disaggregation, repeatability, and universality of application. Geographers need to develop methods for delineating space that either resist change over time or accommodate temporal changes smoothly.

A continuing example of delineating space that has important political implications is the process of defining American congressional districts once each decade based on the population census. The need for fair representation, relative uniformity in population numbers in each district, recognition of traditional communities, and accommodation of changing population distributions comprise some of the criteria that need not equate to partisan politics in constructing at least the first approximation of redrawn district boundaries (Monmonier 2001). Some states have nonpartisan commissions to delineate the districts, yet geography provides very little substantive advice on the subject to guide such groups.

Why Do People, Resources, and Ideas Move?

One of the fundamental concepts in geography is the understanding that goods, services, people, energy, materials, money, and even ideas flow through networks and across space from place to place. Although geography faces questions about all these movements, one of the most pressing questions concerns the movement of people. We have some knowledge about the behavior of people who move their residences from one place to another, and we can observe obvious economic forces leading to the migration of people toward locations of relative economic prosperity. However, we have much less understanding about the episodic movements of people in cities. In most developed countries, the congestion of vehicular traffic has become a significant negative feature in assessing the quality of life, and in lesser-developed countries the increasing number of vehicles used in the context of inadequate road networks results in frustrating delays. Geography can and should address fundamental issues such as the environmental consequences of the decision to undertake laborious journeys to work (e.g., contributions of vehicle exhaust to air pollution, the possible environmental changes induced by telecommuting, and the need for alternative-fuel, low-pollution vehicles). In addition to understanding the environmental consequences of daily moves, the discipline has much to offer in describing, explaining, and predicting the sociocultural consequences of these decisions.

The flow of vehicles on roadways involves obvious physical networks, but there are other flows demanding attention that operate through more abstract spaces. The diffusion of culture—particularly "Western" culture, with its emphasis on materialism and individualism—is one of the leading edges of globalization of the world economy. Geographers must begin to address how these social, cultural, and economic forces operate together to diffuse, from a few limited sources, an extensive array of ideas and attitudes that are accepted by a diverse set of receiving populations. Even if such diffusion takes place through digital space, it probably does so in a distinctive geography that we should understand if we are to explain and predict the world in the twenty-first century.

The electrical energy crisis of 2001 made us aware, quite vividly, of the finiteness of nonrenewable resources such as oil and gas and of the difficulties in their distributions. We have already consumed more than 50 percent of the world's known reserves of these resources. Historically, as one energy source has replaced another (as when coal power replaced water power), there have been changes in the locational patterns, growth, importance of settlements, and significance of regions. Examples include the decline of heavy industrial areas into "rust belts" and their replacement with service- and information-based centers that have more locational flexibility. As current energy sources change, what will happen to urban location and growth? Will the geopolitical power structure of the world change markedly? For example, will the countries that are part of the Organization of the Petroleum Exporting Countries (OPEC) retain their global economic power and political strength? Will existing populations and settlements decline, or relocate to alternative sources of energy? What will be the geographic configuration of the economic and political power that goes with such changes?

Finally, the more physically oriented flows, such as those of energy and materials, present a demanding set of questions for geographers. While geochemists are deriving the magnitudes of elemental fluxes of such substances as carbon and nitrogen, for example, it is incumbent on geographers to point out that these fluxes do not take place in aspatial abstract ways, but rather in a physically and socially defined landscape that has important locational characteristics. In other words, although there may very well be an understanding of the amounts of nitrogen circulating from earth to oceans to atmosphere, that circulation is not everywhere equal. How does human management affect the nitrogen and other elemental cycles? What explains its geographic variability? How does that variability change in response to controls not related to human intervention? This leads to our next big question.

How Has the Earth Been Transformed by Human Action?

Humans have altered the earth, its atmosphere, and its water on scales ranging from local to global (Thomas 1956; Turner et al. 1990). At the local scale, many cities and agricultural landscapes represent nearly complete artificiality in a drive to create comfortable places in which to live and work, and to maximize agricultural production for human benefit. The transformations have also had negative effects at local scales, such as altering the chemical characteristics of air and water, converting them into media that are toxic for humans as well as other species. At regional scales, human activities have resulted in

wholesale changes in ecosystems, such as the deforestation of northwest Europe over the past several centuries, a process that seems to be being replicated in many tropical regions today. At a global scale, the introduction of industrial gases into the atmosphere plays a still emerging role in global climate change. Taken together, these transformations have had a geographically variable effect that geographers must better define and explain. Dilsaver and Colten (1992, 9) succinctly outlined the basic questions almost a decade ago: How have human pursuits transformed the environment, and how have human social organizations exerted their control over environments? Graf (2001) recently asked how we can undo some of our previous efforts at environmental change and control.

In many instances, this explanation of variation might emphasize the physical aspects of changes, or understanding the underlying dynamics of why the changes occur (Dilsaver, Wyckoff, and Preston 2000). Wide-ranging assessments of river basins, for example, must rely on a plethora of controlling factors ranging from land use to water, sediment, and contaminant movements. Geographers must employ more complicated and insightful approaches, however, to truly understand why transformations vary from place to place, largely in response to the connection between the biophysical environment and the human society that occupies it. Understanding this delicate interplay between nature and society as a two-way connection can lead us to new knowledge about social and environmental landscapes, but it can also help us make better decisions on how to achieve future landscapes that are more often transformed in nondestructive ways.

One of the primary issues facing many societies in their relationship with their supporting environments is how much of the biophysical world should be left unchanged, or at least changed to the minimal degree possible. The amount of remaining "natural" landscape in many nations is small—probably less than 5 percent of the total surface—so time is growing short to decide what areas should be set aside and preserved (Figure 2).

Figure 2 A local example of transformations brought about in the natural world by humans. The lower Sandy River of western Oregon appears to be a pristine river, but it has radically altered water, sediment, and biological systems because of upstream dams. Photo by W. L. Graf.

Not only do these preservation decisions affect land and water surfaces; they also profoundly affect nonhuman species that use the surfaces for habitat. If human experience is enriched by diverse ecosystems, then the decline in biodiversity impoverishes humanity as well. Which areas should be preserved and why? How should preserved areas be linked with one another? How can public and private property productively coexist with nearby preserved areas?

What Role Will Virtual Systems Play in Learning about the World?

Stated another way, what will virtual systems allow us to do in the future that we cannot do now? What new problems can be pursued (Golledge forthcoming)? Providing an answer opens a Pandora's box of questions concerning the geographic impacts of new technologies (Goodchild 2000). What new multimodal interfaces for interpreting visualized onscreen data need to be developed in order to overcome current technological constraints of geographic data visualization? Can we produce a virtual geography? Do we really want to?

One serious problem that deserves immediate attention is the examination of the geographic implications of the development of economies and societies based on information technology. In particular, the sociospatial implications of an increasing division between the digital haves and have-nots demand attention. Pursuing such a problem will require answering questions about the geographic consequences of employment in cyberspace and its implication for human movements such as migration, intraurban mobility, commuting, and activity-space restructuring.

The current extensive demand for and use of transportation for business purposes may need to be re-examined. It may be argued that, in the world of business communication, geographic distance is a decreasingly important factor, because both digital and visual interaction can take place at the click of a mouse button without the need for person-to-person confrontation. If this is so, what are the longer-term impacts for living and lifestyles, and how could the inhabitation and use of geographic environments be affected? If this is true, why is it that we see dramatic concentrations of cyber-businesses in a few areas, similar to the locational behavior of pre-digital industries? Are Silicon Valley in California and Route 128 in Massachusetts simply the "rust belts" of the future?

Research has shown that the most effective way of learning about an environment is by directly experiencing it, so that all sensory modalities are activated during that experience (Figure 3) (Gale 1984; Lloyd and Heivly 1987; MacEachren 1992). However, many places are distant or inaccessible to most people. The interior of the Amazon rainforest, the arctic tundra of northern Siberia, Himalayan peaks, the interior of the Sahara desert, Antarctica and the South Pole, the barrios of Rio de Janeiro, and the Bosnian highlands can become much closer to us. Satellite imagery provides detailed digitized imagery of these places. A problem awaiting solution is how to use this extensive digital database to build virtual systems that will allow immersive experiences with such environments. Problems of motion-sickness experienced by some people in immersive systems need to be solved; assuming this will be achieved, virtual reality could become the laboratory of the future for experiencing different places and regions around the world.

Discovering how best to deal with problematic futures, on earth or on other planets, is definitely one of the big problems

Figure 3 Exploring immersive virtual worlds with equipment developed between 1992 and 2001, showing the original and the miniaturized versions of a GPS-driven auditory virtual environment at the University of California, Santa Barbara. The more cumbersome 1992 version is shown on the left, with the reduced 2001 version on the right. Psychologist Jack Loomis and associates developed the system, demonstrated here by author Reginald Golledge. Photo courtesy of R. Golledge.

facing current and future geographers. Many land use planning, transportation, and social policies are made on an "if then" basis. Because we are unable to change the world experimentally, we need to investigate other ways of observing environmental events and changes. Examples include changing a street for vehicles to a pedestrian mall to explore human movement behavior, or experiencing the action and consequences of snow or mud avalanches in tourist-dependent alpine environments. What more can we learn by building and manipulating virtual environments? In a virtual system, we can raise local pollution levels, accelerate global warming, change sea levels by melting ice caps, or simulate the impacts of strictly enforcing land conversion policies at the rural-urban fringes of large cities. In the face of an increasingly international economy and globalization of environmental issues, there is a need to develop a way to explore possible scenarios before implementing policies theoretically designed to deal with global (or more local) problems.

How Do We Measure the Unmeasurable?

Geography is normally practiced at local to national scales at which we can get a clear sense of the existence or development of patterns and processes. People, landscapes, and resources are not evenly distributed on the earth's surface, so we begin with a palette that is diverse. How can we accommodate such diversity in policies to avoid winners and losers? Economists, for example, assume away all spatial variability in their economic models. What happens to general models when space is introduced? How can we transform from the local to the global and vice versa? The question of scale transformation, especially the calibration of large-scale global circulation models or the development of climate-impact models globally with local or regional applications is a major area in which geography can contribute and is playing a leading role (AAG GCLP Group forthcoming).

We need to develop more compatible databases that have an explicit geographic component, with geocoded data that permit us to scale up and scale down as the need arises. Data collection, archiving, and dissemination all are areas that require our expertise, be it demographic data, environmental data, or land-use data. The large question is, how do we maintain a global information system that goes beyond the petty tyrannies of nation-states (and the need to protect information for "security" reasons), yet protects individuals' right to privacy? The selective use of remote-sensing techniques to monitor environmental conditions has been helpful in understanding the linkages between local activities and global impacts. However, can we use advanced technology to support demographic data collection and analyses and still maintain safeguards on privacy protection (Liverman et al. 1998)? For example, recent Supreme Court decisions have placed important legal protections on the use of thermal infrared sensors in public safety.

Another series of issues involves the aggregation of human behaviors. How can we geographically aggregate data along a set of common dimensions to insure its representation of reality

and get around the thorny issues of averaging and the mean-areal-center or modified areal-unit problems? We often use techniques to handle aggregated populations and areas that in fact, depart from reality, creating a type of artificial environment. Unfortunately, public policies all too often are based on these constructed realities, thus further exacerbating the distribution of goods, services, and resources. What new spatial statistical tools do we need to address this concern?

Lastly, in a post-11-September world, how do we measure the geography of fear? Does the restriction of geographic data (presumably for national security reasons) attenuate or amplify fear of the unknown? The discipline requires the open access to information and data about the world and the people who live there. Data access will be one of the key issues for our community to address in the coming years.

What Role Has Geographical Skill Played in the Evolution of Human Civilization, and What Role Can It Play in Predicting the Future?

Is there a necessary geographic base to human history? If so, how can we improve our ability to predict spatial events and events that have spatial consequences that will fundamentally shape the future? Can we develop the geographic equivalent of leading economic indicators?

From the early cradles of civilization in Africa and Asia, humankind gradually colonized the earth. This process of redistributing people in space (migration) was caused by population growth, resource exhaustion, attractive untapped resources, environmental change, environmental hazard, disease, or invasion and succession by other human groups. But what skills and abilities were required to ensure success in relocation movements? Were the movements random or consciously directed? If they were directed, then what skills and/or abilities were required by explorers, leaders, and followers to ensure success? What criteria had to be satisfied before resettlement was possible? What new geographic skills and abilities have been developed throughout human history, and which ones have deteriorated or disappeared? Have geographic skills and abilities been maintained equally in males and females? If not, what developments in the evolution of human civilizations have mediated such losses or changes?

While we know much about human history, we know little about the geographical basis of world history, and we know little of the extent to which the presence or absence of geographic knowledge played a significant part in historical development. For example, would Napoleon's invasion of Russia been more successful had skilled and knowledgeable geographers counseled him on the route chosen and the appropriate season for movement? Historians often tell us that understanding the past is the key to knowing the present and to successfully predicting the future. We cannot fully understand the past if we ignore or diminish the importance of environmental diversity and knowledge about those variations that are the result of spatial and geographic thinking and reasoning. A similar argument can be made for predicting future events and behaviors. What geographic knowledge is likely to be important in prediction? Must we rely on assumptions about uniform environments, population characteristics, tastes and preferences, customs, beliefs, and values? Such a procedure is precarious at best. However, we do not currently know how to incorporate geographic variability into our models, or indeed what variables should be incorporated into predictive models. Achieving such a goal is a necessary part of increasing our very limited predictive capabilities.

How and Why Do Sustainability and Vulnerability Change from Place to Place and over Time?

Historically, geography was an integrative science with a particular focus on regions. It then switched from breadth to depth, with improvements in theory, methods, and techniques. We are now returning to that earlier perspective as we look for common ground in the interactions between human systems and physical systems. Increasing population pressures, the regional depletion or total exhaustion of resources, environmental degradation, and rampant development are processes that affect the sustainability of natural systems and constructed environments. There is a movement toward the integration of many different social and natural science perspectives into a field called sustainability science (Kates et al. 2001). Understanding what constrains and enhances sustainable environments will be an important research theme in the future. How can we maintain and improve the quality of urban environments for general living (social, economic, and environmental conditions)? How long can the processes of urban and suburban growth continue without deleterious and fundamental changes in the landscape and the escalation in costs of environmental restoration? Suburban sprawl is already a major policy issue. What is the long-term impact on human survival of the constant usurpation of agricultural land by the built environment? How long can we continue slash and burn agriculture in many parts of the tropical world? What triggers the environmental insecurity of nations, and how does this lead to armed conflicts and mass migrations of people? How have these processes varied in time and space? What are the greatest threats to the sustainability of human settlements, agriculture, energy use, for example and how can we mitigate or reduce those threats (NRC 2000)?

Nonsustainable environments enhance the effect of risks and hazards and ultimately increase both biophysical and social vulnerability, often resulting in disasters of one kind or another. When societies or ecosystems lack the ability to stop decay or decline and they do not have the adequate means to defend against such changes, there can be potentially catastrophic results. Examples include the environmental degradation of the Aral Sea, the increasing AIDS pandemic, and the human and environmental costs of coastal living (Heinz Center 2000). Vulnerability can be thought of as a continuum of processes,

ranging from the initial susceptibility to harm to resilience (the ability to recover) to longer-term adaptations in response to large-scale environmental changes (Cutter, Mitchell, and Scott 2000). These processes manifest themselves at different geographic scales, ranging from the local to the global. What is the threshold when vulnerability ceases to become something we can deal with and becomes something we cannot? At what point does the built environment or ecosystem extend beyond its own ability to recover from natural or social forces?

What Is the Nature of Spatial Thinking, Reasoning, and Abilities?

Geographic knowledge is the product of spatial thinking and reasoning (Golledge 2002b). These processes require the ability to comprehend scale changes; transformations of phenomena, or representations among one, two, and three spatial dimensions. They also require understanding of: the effect of distance, direction, and orientation on developing spatial knowledge; the nature of reference frames for identifying locations, distributions and patterns; the nature of spatial hierarchies; the nature of forms by extrapolating from cross sections; the significance of adjacency and nearest neighbor concepts; the spatial properties of density, distance, and density decay; and the configurations of patterns and shapes in various dimensions and with differing degrees of completeness. It also requires knowing the implications of spatial association and understanding other concepts not yet adequately articulated or understood. What geography currently lacks is an elaboration of the fundamental geographic concepts and skills that are necessary for the production and communication of spatial and geographic information. In the long run, this will be needed before geography can develop a well-articulated knowledge base of a type similar to other human and physical sciences.

Conclusion

In the American Declaration of Independence, Thomas Jefferson wrote that among the most basic of human rights are life, liberty, and the pursuit of happiness. Each of these rights is played out upon a geographic stage, has geographic properties, and operates as a geographical process. Geography, as a field of knowledge and as a perspective on the world, has paid too little attention to these grand ideas, and they are fertile ground for the seeds of new geographic research. How and why does the opportunity for the pursuit of happiness vary from one place to another, and does the very nature of that pursuit change geographically?

In pursuit of answers to the big questions articulated above, we will inevitably need to think about doing research on problems such as:

- What are the spatial constraints on pursuing goals of life, liberty, and the pursuit of happiness?
- What are our future resource needs, and where will we find the new resources that have not, at this stage, been adequately explored?

- When does geography start and finish? Does it matter?
- What are likely to be the major problems in doing the geography of other planets?
- Will cities of the future remain bound to the land surface, or will they move to what we now consider unlikely or exotic locations (under water or floating in space)?

The big questions posed here are not all encompassing. They represent our collective judgments (and biases) on what issues are significant for the discipline, and those that should provide a focus for our considerable intellectual capital. Not everyone will agree with us, nor should they. We view this article as the beginning of a dialogue within the discipline as to what are the probable big questions for the next generation of geographers.

References

Abler, R. F. 2001. From the meridian—Wilford's "science writer's view of geography." *AAG Newsletter* 36 (4):2, 9.

Association of American Geographers (AAG) Global Change in Local Places (GCLP) Research Group. Forthcoming. *Global change and local places: Estimating, understanding, and reducing greenhouse gases.* Cambridge, U.K.: Cambridge University Press.

Cutter, S. L., J. T. Mitchell, and M. S. Scott. 2000. Revealing the vulnerability of people and places: A case study of Georgetown County, South Carolina. *Annals of the Association of American Geographers* 90:713–37.

Dilsaver, L. M., and C. E. Colten, eds. 1992. *The American environment: Interpretations of past geographies.* Lanham, MD: Rowan and Littlefield Publishers.

Dilsaver, L. M., W. Wyckoff, and W. L. Preston. 2000. Fifteen events that have shaped California's human landscape. *The California Geographer* XL:1–76.

Gale, N. D. 1984. Route learning by children in real and simulated environments. Ph.D. diss., Department of Geography, University of California, Santa Barbara.

Golledge, R. G. 2002. The nature of geographic knowledge. *Annals of the Association of American Geographers* 92 (1):1–14.

———. Forthcoming. *Spatial cognition and converging technologies.* Paper presented at the Workshop on Converging Technology (NBIC) for Improving Human Performance, sponsored by the National Science Foundation. Washington, D.C. In press.

Goodchild, M. F. 2000. Communicating geographic information in a digital age. *Annals of the Association of American Geographers* 90:344–55.

Graf, W. L. 2001. Damage control: Restoring the physical integrity of America's rivers. *Annals of the Association of American Geographers* 91:1–27.

Heinz Center. 2000. *The hidden costs of coastal erosion.* Washington, D.C.: The H. John Heinz III Center for Science, Economics and the Environment.

Kates, R. W., W. C. Clark, R. Corell, J. M. Hall, C. C. Jaeger, I. Lowe, J. J. McCarthy, H. J. Schnellnhuber, B. Bolin, N. M. Dickson, S. Faucheux, G. C. Gallopin, A. Grubler, B. Huntley, J. Jager, N. S. Jodha, R. E. Kasperson, A. Mabogunje, P. Matson,

H. Mooney, B. Moore III, T. O'Riodan, and U. Svedin. 2001. Sustainability science. *Science* 292:641–42.

Liverman, D., E. F. Moran, R. R. Rindfuss, and P. C. Stern, eds. 1998. *People and pixels: Linking remote sensing and social science.* Washington, D.C.: National Academy Press.

Lloyd, R. E., and C. Heivly. 1987. Systematic distortion in urban cognitive maps. *Annals of the Association of American Geographers* 77:191–207.

MacEachren, A. M. 1992. Application of environmental learning theory to spatial knowledge acquisition from maps. *Annals of the Association of American Geographers* 82 (2):245–74.

Monmonier, M. S. 2001. *Bushmanders and Bullwinkles: How politicians manipulate electronic maps and census data to win elections.* Chicago: University of Chicago Press.

Montello, D. R. 1993. Scale and multiple psychologies of space. In *Spatial information theory: A theoretical basis for GIS. Lecture notes in computer science 716. Proceedings, European Conference, COSIT '93. Marciana Marina, Elba Island, Italy, September,* ed. A. U. Frank and I. Campari. 312–21. New York: Springer-Verlag.

National Research Council (NRC). 1999. *New strategies for America's watersheds.* Washington, D.C.: National Academy Press.

———. 2000. *Our common journey: A transition toward sustainability.* Washington, D.C.: National Academy Press.

Thomas, W. L., Jr., ed. 1956. *Man's role in changing the face of the earth.* Chicago: The University of Chicago Press.

Turner, B. L. II, W. C. Clark, R. W. Kates, J. F. Richards, J. T. Mathews, and W. Meyer, eds. 1990. *The earth as transformed by human action: Global and regional changes in the biosphere over the past 300 years.* Cambridge, U.K.: University of Cambridge Press.

Worster, D. E. 1979. *Dust bowl: The Southern plains in the 1930s.* Oxford: Oxford University Press.

———. 1993. *The wealth of nature: Environmental history and the ecological imagination.* New York: Oxford University Press.

SUSAN L. CUTTER is Carolina Distinguished Professor, Department of Geography, University of South Carolina, Columbia, SC 29208. E-mail: scutter@sc.edu. She served as president of the Association of American Geographers from 2000–2001, and is a fellow of the American Association for the Advancement of Science (AAAS). Her research interests are vulnerability science, and environmental hazards policy and management. **REGINALD GOLLEDGE** is a Professor of Geography at the University of California, Santa Barbara, Santa Barbara, CA 93106. E-mail: golledge@geog.ucsb.edu and served as AAG president from 1999 to 2000. His research interests include various aspects of behavioral geography (spatial cognition, cognitive mapping, spatial thinking), the geography of disability (particularly the blind), and the development of technology (guidance systems and computer interfaces) for blind users. **WILLIAM L. GRAF** is Education Foundation University Professor and Professor of Geography at the University of South Carolina, Columbia, SC 29208. E-mail: graf@sc.edu. He served as AAG president from 1998–1999, and is a National Associate of the National Academy of Science. His specialties are fluvial geomorphology and policy for public land and water.

From *The Professional Geographer,* August 2002, pp. 305–317. Copyright © 2002 by Taylor & Francis Journals. Reprinted by permission.

Point of View

Rediscovering the Importance of Geography

Alexander B. Murphy

As Americans struggle to understand their place in a world characterized by instant global communications, shifting geopolitical relationships, and growing evidence of environmental change, it is not surprising that the venerable discipline of geography is experiencing a renaissance in the United States. More elementary and secondary schools now require courses in geography, and the College Board is adding the subject to its Advanced Placement program. In higher education, students are enrolling in geography courses in unprecedented numbers. Between 1985–86 and 1994–95, the number of bachelor's degrees awarded in geography increased from 3,056 to 4,295. Not coincidentally, more businesses are looking for employees with expertise in geographical analysis, to help them analyze possible new markets or environmental issues.

In light of these developments, institutions of higher education cannot afford simply to ignore geography, as some of them have, or to assume that existing programs are adequate. College administrators should recognize the academic and practical advantages of enhancing their offerings in geography, particularly if they are going to meet the demand for more and better geography instruction in primary and secondary schools. We cannot afford to know so little about the other countries and peoples with which we now interact with such frequency, or about the dramatic environmental changes unfolding around us.

From the 1960s through the 1980s, most academics in the United States considered geography a marginal discipline, although it remained a core subject in most other countries. The familiar academic divide in the United States between the physical sciences, on one hand, and the social sciences and humanities, on the other, left little room for a discipline concerned with how things are organized and relate to one another on the surface of the earth—a concern that necessarily bridges the physical and cultural spheres. Moreover, beginning in the 1960s, the U.S. social-science agenda came to be dominated by pursuit of more-scientific explanations for human phenomena, based on assumptions about global similarities in human institutions, motivations, and actions. Accordingly, regional differences often were seen as idiosyncrasies of declining significance.

Although academic administrators and scholars in other disciplines might have marginalized geography, they could not kill it, for any attempt to make sense of the world must be based on some understanding of the changing human and physical patterns that shape its evolution—be they shifting vegetation zones or expanding economic contacts across international boundaries. Hence, some U.S. colleges and universities continued to teach geography, and the discipline was often in the background of many policy issues—for example, the need to assess the risks associated with foreign investment in various parts of the world.

By the late 1980s, Americans' general ignorance of geography had become too widespread to ignore. Newspapers regularly published reports of surveys demonstrating that many Americans could not identify major countries or oceans on a map. The real problem, of course, was not the inability to answer simple questions that might be asked on *Jeopardy!;* instead, it was what that inability demonstrated about our collective understanding of the globe.

Geography's renaissance in the United States is due to the growing recognition that physical and human processes such as soil erosion and ethnic unrest are inextricably tied to their geographical context. To understand modern Iraq, it is not enough to know who is in power and how the political system functions. We also need to know something about the country's ethnic groups and their settlement patterns, the different physical environments and resources within the country, and its ties to surrounding countries and trading partners.

Those matters are sometimes addressed by practitioners of other disciplines, of course, but they are rarely central to the analysis. Instead, generalizations are often made at the level of the state, and little attention is given to spatial patterns and practices that play out on local levels or across international boundaries. Such preoccupations help to explain why many scholars were caught off guard by the explosion of ethnic unrest in Eastern Europe following the fall of the Iron Curtain.

Similarly, comprehending the dynamics of El Niño requires more than knowledge of the behavior of ocean and air currents;

it is also important to understand how those currents are situated with respect to land masses and how they relate to other climatic patterns, some of which have been altered by the burning of fossil fuels and other human activities. And any attempt to understand the nature and extent of humans' impact on the environment requires consideration of the relationship between human and physical contributions to environmental change. The factories and cars in a city produce smog, but surrounding mountains may trap it, increasing air pollution significantly.

Today, academics in fields including history, economics, and conservation biology are turning to geographers for help with some of their concerns. Paul Krugman, a noted economist at the Massachusetts Institute of Technology, for example, has turned conventional wisdom on its head by pointing out the role of historically rooted regional inequities in how international trade is structured.

Geographers work on issues ranging from climate change to ethnic conflict to urban sprawl. What unites their work is its focus on the shifting organization and character of the earth's surface. Geographers examine changing patterns of vegetation to study global warming; they analyze where ethnic groups live in Bosnia to help understand the pros and cons of competing administrative solutions to the civil war there; they map AIDS cases in Africa to learn how to reduce the spread of the disease.

Geography is reclaiming attention because it addresses such questions in their relevant spatial and environmental contexts. A growing number of scholars in other disciplines are realizing that it is a mistake to treat all places as if they were essentially the same (think of the assumptions in most economic models), or to undertake research on the environment that does not include consideration of the relationships between human and physical processes in particular regions.

Still, the challenges to the discipline are great. Only a small number of primary- and secondary-school teachers have enough training in geography to offer students an exciting introduction to the subject. At the college level, many geography departments are small; they are absent altogether at some high-profile universities.

Perhaps the greatest challenge is to overcome the public's view of geography as a simple exercise in place-name recognition. Much of geography's power lies in the insights it sheds on the nature and meaning of the evolving spatial arrangements and landscapes that make up our world. The importance of those insights should not be underestimated at a time of changing political boundaries, accelerated human alteration of the environment, and rapidly shifting patterns of human interaction.

ALEXANDER B. MURPHY is a professor and head of the geography department at the University of Oregon, and a vice-president of the American Geographical Society.

From *Chronicle of Higher Education* by Alexander B. Murphy, October 30, 1998. Copyright © 1998 by Alexander B. Murphy. Reproduced with permission of the author.

The Four Traditions of Geography

WILLIAM D. PATTISON

Late Summer, 1990

To Readers of the *Journal of Geography:*

I am honored to be introducing, for a return to the pages of the *Journal* after more than 25 years, "The Four Traditions of Geography," an article which circulated widely, in this country and others, long after its initial appearance—in reprint, in xerographic copy, and in translation. A second round of life at a level of general interest even approaching that of the first may be too much to expect, but I want you to know in any event that I presented the paper in the beginning as my gift to the geographic community, not as a personal property, and that I re-offer it now in the same spirit.

In my judgment, the article continues to deserve serious attention—perhaps especially so, let me add, among persons aware of the specific problem it was intended to resolve. The background for the paper was my experience as first director of the High School Geography Project (1961–63)—not all of that experience but only the part that found me listening, during numerous conference sessions and associated interviews, to academic geographers as they responded to the project's invitation to locate "basic ideas" representative of them all. I came away with the conclusion that I had been witnessing not a search for consensus but rather a blind struggle for supremacy among honest persons of contrary intellectual commitment. In their dialogue, two or more different terms had been used, often unknowingly, with a single reference, and no less disturbingly, a single term had been used, again often unknowingly, with two or more different references. The article was my attempt to stabilize the discourse. I was proposing a basic nomenclature (with explicitly associated ideas) that would, I trusted, permit the development of mutual comprehension **and** confront all parties concerned with the pluralism inherent in geographic thought.

This intention alone could not have justified my turning to the NCGE as a forum, of course. The fact is that from the onset of my discomfiting realization I had looked forward to larger consequences of a kind consistent with NCGE goals. As finally formulated, my wish was that the article would serve "to greatly expedite the task of maintaining an alliance between professional geography and pedagogical geography and at the same time to promote communication with laymen" (see my fourth paragraph). I must tell you that I have doubts, in 1990, about the acceptability of my word choice, in saying "professional," "pedagogical," and "layman" in this context, but the message otherwise is as expressive of my hope now as it was then.

I can report to you that twice since its appearance in the *Journal,* my interpretation has received more or less official acceptance—both times, as it happens, at the expense of the earth science tradition. The first occasion was Edward Taaffe's delivery of his presidential address at the 1973 meeting of the Association of American Geographers (see *Annals AAG,* March 1974, pp. 1–16). Taaffe's working-through of aspects of an interrelation among the spatial, area studies, and man-land traditions is by far the most thoughtful and thorough of any of which I am aware. Rather than fault him for omission of the fourth tradition, I compliment him on the grace with which he set it aside in conformity to a meta-epistemology of the American university which decrees the integrity of the social sciences as a consortium in their own right. He was sacrificing such holistic claims as geography might be able to muster for a freedom to argue the case for geography as a social science.

The second occasion was the publication in 1984 of *Guidelines for Geographic Education: Elementary and Secondary Schools,* authored by a committee jointly representing the AAG and the NCGE. Thanks to a recently published letter (see *Journal of Geography,* March–April 1990, pp. 85–86), we know that, of five themes commended to teachers in this source,

> The committee lifted the human environmental interaction theme directly from Pattison. The themes of place and location are based on Pattison's spatial or geometric geography, and the theme of region comes from Pattison's area studies or regional geography.

Having thus drawn on my spatial, area studies, and man-land traditions for four of the five themes, the committee could have found the remaining theme, movement, there too—in the spatial tradition (see my sixth paragraph). However that may be, they did not avail themselves of the earth science tradition, their reasons being readily surmised. Peculiar to the elementary and secondary schools is a curriculum category framed as much by theory of citizenship as by theory of knowledge: the social studies. With admiration, I see already in the committee members' adoption of the theme idea a strategy for assimilation of their program to the established repertoire of social studies practice. I see in their exclusion of the earth science tradition an intelligent respect for social studies' purpose.

Here's to the future of education in geography: may it prosper as never before.

W. D. P., 1990

Reprinted from the Journal of Geography, 1964, pp. 211–216

In 1905, one year after professional geography in this country achieved full social identity through the founding of the Association of American Geographers, William Morris Davis responded to a familiar suspicion that geography is simply an undisciplined "omnium-gatherum" by describing an approach that as he saw it imparts a "geographical quality" to some knowledge and accounts for the absence of the quality elsewhere.[1] Davis spoke as president of the AAG. He set an example that was followed by more than one president of that organization. An enduring official concern led the AAG to publish, in 1939 and in 1959, monographs exclusively devoted to a critical review of definitions and their implications.[2]

Every one of the well-known definitions of geography advanced since the founding of the AAG has had its measure of success. Tending to displace one another by turns, each definition has said something true of geography.[3] But from the vantage point of 1964, one can see that each one has also failed. All of them adopted in one way or another a monistic view, a singleness of preference, certain to omit if not to alienate numerous professionals who were in good conscience continuing to participate creatively in the broad geographic enterprise.

The thesis of the present paper is that the work of American geographers, although not conforming to the restrictions implied by any one of these definitions, has exhibited a broad consistency, and that this essential unity has been attributable to a small number of distinct but affiliated traditions, operant as binders in the minds of members of the profession. These traditions are all of great age and have passed into American geography as parts of a general legacy of Western thought. They are shared today by geographers of other nations.

There are four traditions whose identification provides an alternative to the competing monistic definitions that have been the geographer's lot. The resulting pluralistic basis for judgment promises, by full accommodation of what geographers do and by plain-spoken representation thereof, to greatly expedite the task of maintaining an alliance between professional geography and pedagogical geography and at the same time to promote communication with laymen. The following discussion treats the traditions in this order: (1) a spatial tradition, (2) an area studies tradition, (3) a man-land tradition and (4) an earth science tradition.

Spatial Tradition

Entrenched in Western thought is a belief in the importance of spatial analysis, of the act of separating from the happenings of experience such aspects as distance, form, direction and position. It was not until the 17th century that philosophers concentrated attention on these aspects by asking whether or not they were properties of things-in-themselves. Later, when the 18th century writings of Immanuel Kant had become generally circulated, the notion of space as a category including all of these aspects came into widespread use. However, it is evident that particular spatial questions were the subject of highly organized answering attempts long before the time of any of these cogitations. To confirm this point, one need only be reminded of the compilation of elaborate records concerning the location of things in ancient Greece. These were records of sailing distances, of coastlines and of landmarks that grew until they formed the raw material for the great *Geographia* of Claudius Ptolemy in the 2nd century A.D.

A review of American professional geography from the time of its formal organization shows that the spatial tradition of thought had made a deep penetration from the very beginning. For Davis, for Henry Gannett and for most if not all of the 44 other men of the original AAG, the determination and display of spatial aspects of reality through mapping were of undoubted importance, whether contemporary definitions of geography happened to acknowledge this fact or not. One can go further and, by probing beneath the art of mapping, recognize in the behavior of geographers of that time an active interest in the true essentials of the spatial tradition—*geometry* and *movement*. One can trace a basic favoring of movement as a subject of study from the turn-of-the-century work of Emory R. Johnson, writing as professor of transportation at the University of Pennsylvania, through the highly influential theoretical and substantive work of Edward L. Ullman during the past 20 years and thence to an article by a younger geographer on railroad freight traffic in the U.S. and Canada in the *Annals* of the AAG for September 1963.[4]

One can trace a deep attachment to geometry, or positioning-and-layout, from articles on boundaries and population densities in early 20th century volumes of the *Bulletin of the American Geographical Society,* through a controversial pronouncement by Joseph Schaefer in 1953 that granted geographical legitimacy only to studies of spatial patterns[5] and so onward to a recent *Annals* report on electronic scanning of cropland patterns in Pennsylvania.[6]

One might inquire, is discussion of the spatial tradition, after the manner of the remarks just made, likely to bring people within geography closer to an understanding of one another and people outside geography closer to an understanding of geographers? There seem to be at least two reasons for being hopeful. First, an appreciation of this tradition allows one to see a bond of fellowship uniting the elementary school teacher, who attempts the most rudimentary instruction in directions and mapping, with the contemporary research geographer, who dedicates himself to an exploration of central-place theory. One cannot only open the eyes of many teachers to the potentialities of their own instruction, through proper exposition of the spatial tradition, but one can also "hang a bell" on research quantifiers in geography, who are often thought to have wandered so far in their intellectual adventures as to have become lost from the rest. Looking outside geography, one may anticipate benefits from the readiness of countless persons to associate the name "geography" with maps. Latent within this readiness is a willingness to recognize as geography, too, what maps are about—and that is the geometry of and the movement of what is mapped.

Area Studies Tradition

The area studies tradition, like the spatial tradition, is quite strikingly represented in classical antiquity by a practitioner to whose surviving work we can point. He is Strabo, celebrated for his *Geography* which is a massive production addressed to the statesmen of Augustan Rome and intended to sum up and regularize knowledge not of the location of places and associated cartographic facts, as in the somewhat later case of Ptolemy, but of the nature of places, their character and their differentiation. Strabo exhibits interesting attributes of the area-studies tradition that can hardly be overemphasized. They are a pronounced tendency toward subscription primarily to literary standards, an almost omnivorous appetite for information and a self-conscious companionship with history.

It is an extreme good fortune to have in the ranks of modern American geography the scholar Richard Hartshorne, who has pondered the meaning of the area-studies tradition with a legal acuteness that few persons would challenge. In his *Nature of Geography,* his 1939 monograph already cited,[7] he scrutinizes exhaustively the implications of the "interesting attributes" identified in connection with Strabo, even though his concern is with quite other and much later authors, largely German. The major literary problem of unities or wholes he considers from every angle. The Gargantuan appetite for miscellaneous information he accepts and rationalizes. The companionship between area studies and history he clarifies by appraising the so-called idiographic content of both and by affirming the tie of both to what he and Sauer have called "naively given reality."

The area-studies tradition (otherwise known as the chorographic tradition) tended to be excluded from early American professional geography. Today it is beset by certain champions of the spatial tradition who would have one believe that somehow the area-studies way of organizing knowledge is only a subdepartment of spatialism. Still, area-studies as a method of presentation lives and prospers in its own right. One can turn today for reassurance on this score to practically any issue of the *Geographical Review,* just as earlier readers could turn at the opening of the century to that magazine's forerunner.

What is gained by singling out this tradition? It helps toward restoring the faith of many teachers who, being accustomed to administering learning in the area-studies style, have begun to wonder if by doing so they really were keeping in touch with professional geography. (Their doubts are owed all too much to the obscuring effect of technical words attributable to the very professionals who have been intent, ironically, upon protecting that tradition.) Among persons outside the classroom the geographer stands to gain greatly in intelligibility. The title "area-studies" itself carries an understood message in the United States today wherever there is contact with the usages of the academic community. The purpose of characterizing a place, be it neighborhood or nation-state, is readily grasped. Furthermore, recognition of the right of a geographer to be unspecialized may be expected to be forthcoming from people generally, if application for such recognition is made on the merits of this tradition, explicitly.

Man-Land Tradition

That geographers are much given to exploring man-land questions is especially evident to anyone who examines geographic output, not only in this country but also abroad. O. H. K. Spate,

taking an international view, has felt justified by his observations in nominating as the most significant ancient precursor of today's geography neither Ptolemy nor Strabo nor writers typified in their outlook by the geographies of either of these two men, but rather Hippocrates, Greek physician of the 5th century B.C. who left to posterity an extended essay, *On Airs, Waters and Places.*[8] In this work made up of reflections on human health and conditions of external nature, the questions asked are such as to confine thought almost altogether to presumed influence passing from the latter to the former, questions largely about the effects of winds, drinking water and seasonal changes upon man. Understandable though this uni-directional concern may have been for Hippocrates as medical commentator, and defensible as may be the attraction that this same approach held for students of the condition of man for many, many centuries thereafter, one can only regret that this narrowed version of the man-land tradition, combining all too easily with social Darwinism of the late 19th century, practically overpowered American professional geography in the first generation of its history.[9] The premises of this version governed scores of studies by American geographers in interpreting the rise and fall of nations, the strategy of battles and the construction of public improvements. Eventually this special bias, known as environmentalism, came to be confused with the whole of the man-land tradition in the minds of many people. One can see now, looking back to the years after the ascendancy of environmentalism, that although the spatial tradition was asserting itself with varying degrees of forwardness, and that although the area-studies tradition was also making itself felt, perhaps the most interesting chapters in the story of American professional geography were being written by academicians who were reacting against environmentalism while deliberately remaining within the broad man-land tradition. The rise of culture historians during the last 30 years has meant the dropping of a curtain of culture between land and man, through which it is asserted all influence must pass. Furthermore work of both culture historians and other geographers has exhibited a reversal of the direction of the effects in Hippocrates, man appearing as an independent agent, and the land as a sufferer from action. This trend as presented in published research has reached a high point in the collection of papers titled *Man's Role in Changing the Face of the Earth.* Finally, books and articles can be called to mind that have addressed themselves to the most difficult task of all, a balanced tracing out of interaction between man and environment. Some chapters in the book mentioned above undertake just this. In fact the separateness of this approach is discerned only with difficulty in many places; however, its significance as a general research design that rises above environmentalism, while refusing to abandon the man-land tradition, cannot be mistaken.

The NCGE seems to have associated itself with the man-land tradition, from the time of founding to the present day, more than with any other tradition, although all four of the traditions are amply represented in its official magazine, *The Journal of Geography* and in the proceedings of its annual meetings. This apparent preference on the part of the NCGE members *for defining geography in terms of the man-land tradition* is strong evidence of the appeal that man-land ideas, separately stated, have for persons whose main job is teaching. It should be noted, too, that this inclination reflects a proven acceptance by the general public of learning that centers on resource use and conservation.

Earth Science Tradition

The earth science tradition, embracing study of the earth, the waters of the earth, the atmosphere surrounding the earth and the association between earth and sun, confronts one with a paradox. On the one hand one is assured by professional geographers that their participation in this tradition has declined precipitously in the course of the past few decades, while on the other one knows that college departments of geography across the nation rely substantially, for justification of their role in general education, upon curricular content springing directly from this tradition. From all the reasons that combine to account for this state of affairs, one may, by selecting only two, go far toward achieving an understanding of this tradition. First, there is the fact that American college geography, growing out of departments of geology in many crucial instances, was at one time greatly overweighted in favor of earth science, thus rendering the field unusually liable to a sense of loss as better balance came into being. (This one-time disproportion found reciprocate support for many years in the narrowed, environmentalistic interpretation of the man-land tradition.) Second, here alone in earth science does one encounter subject matter in the normal sense of the term as one reviews geographic traditions. The spatial tradition abstracts certain aspects of reality; area studies is distinguished by a point of view; the man-land tradition dwells upon relationships; but earth science is identifiable through concrete objects. Historians, sociologists and other academicians tend not only to accept but also to ask for help from this part of geography. They readily appreciate earth science as something physically associated with their subjects of study, yet generally beyond their competence to treat. From this appreciation comes strength for geography-as-earth-science in the curriculum.

Only by granting full stature to the earth science tradition can one make sense out of the oft-repeated addage, "Geography is the mother of sciences." This is the tradition that emerged in ancient Greece, most clearly in the work of Aristotle, as a wide-ranging study of natural processes in and near the surface of the earth. This is the tradition that was rejuvenated by Varenius in the 17th century as "Geographia Generalis." This is the tradition that has been subjected to subdivision as the development of science has approached the present day, yielding mineralogy, paleontology, glaciology, meterology and other specialized fields of learning.

Readers who are acquainted with American junior high schools may want to make a challenge at this point, being aware that a current revival of earth sciences is being sponsored in those schools by the field of geology. Belatedly, geography has joined in support of this revival.[10] It may be said that in this connection and in others, American professional geography may have faltered in its adherence to the earth science tradition but not given it up.

In describing geography, there would appear to be some advantages attached to isolating this final tradition. Separation improves the geographer's chances of successfully explaining to educators why geography has extreme difficulty in accommodating itself to social studies programs. Again, separate attention allows one to make understanding contact with members of the American public for whom surrounding nature is known as the geographic environment. And finally, specific reference to the geographer's earth science tradition brings into the open the basis of what is, almost without a doubt, morally the most significant concept in the entire geographic heritage, that of the earth as a unity, the single common habitat of man.

An Overview

The four traditions though distinct in logic are joined in action. One can say of geography that it pursues concurrently all four of them. Taking the traditions in varying combinations, the geographer can explain the conventional divisions of the field. Human or cultural geography turns out to consist of the first three traditions applied to human societies; physical geography, it becomes evident, is the fourth tradition prosecuted under constraints from the first and second traditions. Going further, one can uncover the meanings of "systematic geography," "regional geography," "urban geography," "industrial geography," etc.

It is to be hoped that through a widened willingness to conceive of and discuss the field in terms of these traditions, geography will be better able to secure the inner unity and outer intelligibility to which reference was made at the opening of this paper, and that thereby the effectiveness of geography's contribution to American education and to the general American welfare will be appreciably increased.

References

1. William Morris Davis, "An Inductive Study of the Content of Geography," *Bulletin of the American Geographical Society,* Vol. 38, No. 1 (1906), 71.

2. Richard Hartshorne, *The Nature of Geography,* Association of American Geographers (1939), and idem., *Perspective on the Nature of Geography,* Association of American Geographers (1959).

3. The essentials of several of these definitions appear in Barry N. Floyd, "Putting Geography in Its Place," *The Journal of Geography,* Vol. 62, No. 3 (March, 1963), 117–120.

4. William H. Wallace, "Freight Traffic Functions of Anglo-American Railroads," *Annals of the Association of American Geographers,* Vol. 53, No. 3 (September, 1963), 312–331.

5. Fred K. Schaefer, "Exceptionalism in Geography: A Methodological Examination," *Annals of the Association of American Geographers,* Vol. 43, No. 3 (September, 1953), 226–249.

6. James P. Latham, "Methodology for an Instrumental Geographic Analysis," *Annals of the Association of American Geographers,* Vol. 53, No. 2 (June, 1963), 194–209.

7. Hartshorne's 1959 monograph, *Perspective on the Nature of Geography,* was also cited earlier. In this later work, he responds to dissents from geographers whose preferred primary commitment lies outside the area studies tradition.

8. O. H. K. Spate, "Quantity and Quality in Geography," *Annals of the Association of American Geographers,* Vol. 50, No. 4 (December, 1960), 379.

9. Evidence of this dominance may be found in Davis's 1905 declaration: "Any statement is of geographical quality if it contains . . . some relation between an element of inorganic control and one of organic response" (Davis, *loc. cit.*).

10. Geography is represented on both the Steering Committee and Advisory Board of the Earth Science Curriculum Project, potentially the most influential organization acting on behalf of earth science in the schools.

The Changing Landscape of Fear

SUSAN L. CUTTER, DOUGLAS B. RICHARDSON, AND THOMAS J. WILBANKS

In the days following September 11, 2001, all geographers felt a sense of loss—people we knew perished, and along with everyone else we experienced discomfort in our own lives and a diminished level of confidence that the world will be a safe and secure place for our children and grandchildren. Many of us who are geographers felt an urge and a need to see if we could find ways to apply our knowledge and expertise to make the world more secure. A number of our colleagues assisted immediately by sharing specific geographical knowledge (such as Jack Shroder's expert knowledge on the caves in Afghanistan) or more generally by assisting rescue and relief efforts through our technical expertise in Geographic Information System (GIS) and remote sensing (such as Hunter College's Center for the Analysis and Research of Spatial Information and various geographers at federal agencies and in the private sector). Still others sought to enhance the nation's research capacity in the geographical dimensions of terrorism (the Association of American Geographers' Geographical Dimensions of Terrorism project). Many of us have given considerable thought to how our science and practice might be useful in both the short and longer terms. One result is the set of contributions to this book.

But, we fail in our social responsibility if we spend our time thinking of geography as the <u>end</u>. Geography is not the end; it is one of many <u>means</u> to the end. Our concern should be with issues and needs that transcend any one discipline. As we address issues of terrorism, utility without quality is unprofessional, but quality without utility is self-indulgent. Our challenge is to focus not on geography's general importance but on the central issues in addressing terrorism as a new reality in our lives in the United States (although, unfortunately, not a new issue in too many other parts of our world).

The September 11, 2001 events have prompted both immediate and longer-term concerns about the geographical dimensions of terrorism. Potential questions on the very nature of these types of threats, how the public perceives them, individual and societal willingness to reduce vulnerability to such threats, and ultimately our ability to manage their consequences require concerted research on the part of the geographical community, among others. Geographers are well positioned to address some of the initial questions regarding emergency management and response and some of the spatial impacts of the immediate consequences, but the research community is not sufficiently mobilized and networked internally or externally to develop a longer, sustained, and theoretically informed research agenda on the geographical dimensions of terrorism. As noted more than a decade ago, "issues of nuclear war and deterrence [and now terrorism] are inherently geographical, yet our disciplinary literature is either silent on the subject or poorly focused" (Cutter 1988: 132). Recent events provide an opportunity and a context for charting a new path to bring geographical knowledge and skills to the forefront in solving this pressing international problem.

Promoting Landscapes of Fear

Terrorists (and terrorism) seek to exploit the everyday—things that people do, places that they visit, the routines of daily living, and the functioning of institutions. Terrorism is an adaptive threat which changes its target, timing, and mode of delivery as circumstances are altered. The seeming randomness of terrorist attacks (either the work of organized groups or renegade individuals) in both time and space increases public anxiety concerning terrorism. At the most fundamental level, September 11, 2001 was an attack on the two most prominent symbols of U.S. financial and military power: the World Trade Center and the Pentagon (Smith 2001, Harvey 2002). The events represented symbolic victories of chaos over order and normalcy (Alexander 2002), disruptions in and the undermining of global financial markets (Harvey 2002), a nationalization of terror (Smith 2002), and the creation of fear and uncertainty among the public, precisely the desired outcome by the perpetrators. In generating this psychological landscape of fear, people's activity patterns were and are being altered, with widespread social, political, and economic effects. The reduction in air travel by consumers in the weeks and months following September 11, 2001 was but one among many examples.

What Are the Fundamental Issues of Terrorism?

There are a myriad of different ways to identify and examine terrorism issues. Some of these dimensions are quite conventional, others less so. In all cases, geographical understanding provides an essential aspect of the inquiry. There are a number of dimensions of the issues that seem reasonably clear. For instance, one conventional way of looking at the topic is to distinguish four central subject-matter challenges:

1. *Reducing threats,* including a) reducing the reasons why people want to commit terrorist acts, thereby addressing root causes, and b) reducing the ability of potential terrorists to accomplish their aims, or deterrence.

2. *Detecting threats* that have not been avoided, using sensors and signature detection to spot potential actions before they happen and interrupt them.

3. *Reducing vulnerabilities to threats,* focusing on critical sectors and infrastructures, hopefully without sacrificing civil liberties and individual freedoms.

4. *Improving responses to terrorism,* emphasizing "consequence management," and also attributing causation and learning from experience (for example, forensics applied to explosive materials and anthrax strains).

A different way of viewing terrorism is according to time horizons. Immediately after September 11, 2001, governmental leaders told us that the nation was now engaged in a new "war on terrorism" that will last several years, and that our existing knowledge and technologies are needed for this war. Early estimates of the overall U.S. national effort are very large—in the range of $30 to $40 billion per year—including the formation of a new executive department, the Department of Homeland Security. Early priorities include securing national borders, supporting first responders mainly in the Federal Emergency Management Agency (FEMA) and the Department of Justice, defending against bioterrorism, and applying information technologies to improve national security.

Beyond this, we know that better knowledge and practices should be put to use in the next half decade or so, as we face a challenge that is more like a stubborn virus than a single serial killer. To address this type of need, attention often is placed on capabilities where progress can be made relatively quickly if resources are targeted carefully. Some of our CIS and GIScience tools are especially promising candidates for such enhancements, which have both positive and negative consequences (Monmonier 2002). The use of such technologies surely will help secure homelands, but at what price, the loss of personal freedoms or invasion of privacy?

There are other dimensions as well. For instance, one dimension concerns boundaries between free exchanges of information and limited ones, between classified work and unclassified work. Another differentiates between different types of threats: physical violence, chemical or biological agents, cyberterrorism, and the like. Still other themes are woven through the material that follows.

The Challenge Ahead

The greatest challenge to geographers and our colleagues in neighboring fields of study is to stretch our minds beyond familiar research questions and specializations so as to be innovative, even ingenious, in producing new understandings that contribute to increased global security. Clearly, the most

serious specific threats to security in the future will be actions that are difficult to imagine now: social concerns just beginning to bubble to the surface, technologies yet to be developed, biological agents that do not yet exist, terrorist practices that are beyond our imagination. A core challenge is to improve knowledge and institutional capacities that prepare us to deal with the unknown and the unexpected, with constant change calling for staying one step ahead instead of always being one step behind. When research requires, say, three years to produce results and another two years to communicate in print to prospective audiences, we need to be unusually prescient as we construct our research agendas related to terrorism issues, and we need to be very perceptive and skillful in convincing nongeographers that these longer-term research objectives are, in fact, truly important.

The topic of combating terrorism is not an easy one. It calls for us to stretch in directions that may be new and not altogether comfortable. It threatens to entangle us in policy agendas that many of us may consider insensitively conceived, even distasteful. It may endanger social cohesion in our own community of scholars. On the other hand, how can we turn our backs on a phenomenon that threatens political freedom, social cohesion far beyond our own cohorts, economic progress, environmental sustainability, and many other values that we hold dear, including the future security of our own children and grandchildren?

More fundamentally, geographers are not concerned only with winning the war on terrorism in the next two years or deploying new capabilities in the next five or ten. We are concerned with working toward a secure century, restoring a widespread sense of security in the global society in the longer term without undermining basic freedoms. This is the domain of the research world; assuring a stream of new knowledge, understandings, and tools for the longer term, and looking for policies and practices that—if they could be conceived and used—would make a significant difference in the quality of life.

As we prepare to create this new knowledge and understandings, what we are trying to do, in fact, is to create the new twenty-first-century utility—not a hardened infrastructure such as for power or water, but rather a geographical understanding and spatial infrastructure that helps the nation understand and respond to threats. The effort required to create this new utility to serve the nation has an historical analogy in the creation of the Tennessee Valley Authority (TVA), under Franklin Roosevelt's New Deal. The Appalachian region in the southeastern United States had a long history of economic depression and was among those areas hardest hit by the Great Depression of the 1930s. The creation of the TVA, a multipurpose utility with an economic development mission, constructed dams for flood control and hydroelectric power for the region in order to: 1) bring electricity to the rural areas that did not have it; 2) stimulate new industries to promote economic development; 3) control flooding, which routinely plagued the region; and 4) develop a more sustainable and equitable future for the region's residents. This twenty-first-century utility must rely on geographical knowledge and synthesis capabilities as we begin to understand the root causes of insecurity both here and abroad, vulnerabilities and resiliencies in our daily lives and the systems that support them, and our collective role in fostering a more sustainable future, both domestically and globally.

Much of the content of this book is aimed at this longer term, and it is important for geography to join with others in the research community to assure that the long term is not neglected as research support is directed toward combating terrorism and protecting homelands in the short run. This is why the Association of American Geographers and some of its members have joined together to produce the perspectives and insights represented in this book. It is only a start, we still have a long way to go, and there are daunting intellectual and political hazards to be overcome. But if many of us will keep a part of our professional focus on this global and national issue, we have a chance to make our world better in many tangible ways.

The Geography of Ecosystem Services

JAMES BOYD

The study of ecosystem services involves two broad missions. The first is a *biophysical* one associated with ecology, hydrology, and the other natural sciences. How can we protect—or, ideally, enhance—the biophysical goods and services necessary to our well-being? If we want clean air and water, healthy and abundant species populations, pollination, irrigation, protection from floods and fires, how can we take action to preserve these things?

The second is an *economic* mission to measure and communicate the value of those goods and services. Quantitative measures help justify interventions to protect natural resources and systems. They also spur government and other decisionmakers to take ecological gains and losses into account.

Geography is essential to both missions. Ecologists and economists of ecosystem services are scrambling to develop skills in mapping, visualization, and the manipulation of data via geospatial information systems. These skills aren't optional. We eventually need to be able to see and manage what can be called the "missing economy of nature," which is absent for several reasons. In general, markets and business activity do not produce and trade ecosystem goods and services. Consequently, the information we use to measure the conventional economy doesn't capture the free public goods provided by natural systems. Besides, nature is inherently complex. How does an action taken in one place affect conditions in another?

In Nature, Some Things Move, Others Stay Put

From an ecological perspective, geography matters because *nature moves.* Air circulates. Water runs downhill. Species migrate. Seeds and pollen disperse. Not only that, the movement of one thing—say water—tends to trigger the movement of other things, like birds and fish. With the goal of managing and protecting ecosystem goods and services, we must understand this web of movement. You could say that in nature, *nothing stays put.* Ecologically, the constant movement and mixing of natural systems is what generates the need for geographic science.

Interestingly, you could also say that in nature, *everything stays put*—an apparent contradiction. A distinctive feature of ecosystem goods and services—once produced—is that they are unmovable. You can't move a lake, river basin, or forest. You can't ship clean air from one city to another. Birds will migrate where birds migrate. Beautiful mountain trails and scenery can be found in Colorado. Too bad, Kansas. To economists, it is this property of ecosystem goods and services that triggers the need for geography. As any realtor will tell you, three things matter: location, location, location. The same is true for ecosystem goods and services. They're just like houses: if you want to know their value, it's all about the neighborhood.

The Production of Ecosystem Services: Nature in Motion

Think about anything in nature you care about. It could be the beauty of a park, a species you fish for or hunt, or the quality of the air you breathe and water you drink. Now ask the following question: what do those things depend on?

Downstream water quality depends on upstream land uses. The health of Gulf of Mexico fisheries, for example, depends on agricultural practices in the upper Midwest. Air quality in the Adirondacks depends on pollution emissions from the Midwest. Coastal cities and towns depend on nearby wetlands to absorb flood pulses. The point is that the ecosystem goods and services we care about often depend on physical conditions at a great distance from the thing we actually care about. This is a consequence of the continual movement of nature's components.

Accordingly, the biophysical analysis of ecosystem goods and services must be geographic. Treating an ecological problem at the point where it occurs usually doesn't work. It's like putting a band-aid on a lesion caused by an underlying disease. Our ecological diseases—and their cures—are geographic, because ecological systems are geographic.

The challenge for ecosystem scientists and managers is to scientifically relate cause and effect when the cause-and-effect relationship is spatial. We call these relationships *spatial production functions,* because they tell us how an action (good or bad) in one place affects the production of ecosystem goods and services in another. Broadly, we need spatial production functions that describe the dependence of:

- species on the configuration of lands needed for their reproduction, forage, and migration;
- surface and aquifer water volumes and quality on land cover configurations and land uses;

- flood and fire protection services on land cover configurations;
- soil quality on climate variables and land uses; and
- air quality on pollutant emissions, atmospheric processes, and natural sequestration.

The science of these effects is already well underway. For example, we know that stream bank vegetation can improve water quality, help prevent soil erosion, and provide desirable habitat for certain species. But much more remains to be done. We know much less about the exact, empirical relationship between vegetation and water quality.

Why is it such a challenge? First, nature is a highly complex and non-uniform system. Complexity means that causal relationships can only be tested using rigorous, data-intensive empirical and scientific methods that are difficult and costly to perform. Second, non-uniformity means that even if you establish a causal relationship in one location, that relationship may not hold in other locations. Third, empirical analysis of causality requires collaboration between different disciplines (ecology and hydrology, for example). Cross-disciplinary collaboration in any scientific inquiry is always a practical barrier. Finally, the biophysical scientists have many other things to study and have limited financial support for all they are asked to do.

However, deeper understanding of these production functions is necessary if the ecosystem services agenda is to be taken seriously. Ecosystem protection and management will be ineffective at best, and dangerous at worst, if we cannot make credible claims about ecological cause and effect. And the only way to test ecological cause and effect is with spatial—that is, geographic—understanding of biophysical production functions.

The good news is that maps and mapping technology are increasingly capable of capturing and manipulating this data. Water-sheds can now be categorized on the basis of their adjoining land uses. Geographic information system (GIS) tools allow us to "see" migratory pathways and design protections accordingly. As ecology becomes ever more sophisticated in its use of spatial science and data the practical ability to measure cause and effect will become more and more possible.

The Value of Ecosystem Services: Nature's Social Neighborhood

When McDonald's wants to open yet another McDonald's, the first thing the company does is look at a map. Where are the customers? How many competitors are in the vicinity? Do people have easy access from the highway? When economists value ecosystem services, the same kind of things matter. How many people can enjoy the service? Are there other ways to get the service in that neighborhood? Do we have easy access to the service?

Ecosystem goods and services are like houses and fast food outlets because we can't have them shipped to us. They don't move to be near us, we move to be near them. This is most obvious when we talk about recreation. Usually, outdoor recreation requires us to travel to a park, stream, or forest. But backyard ecosystem services are the same. Chances are you chose your house based in part on its proximity to large trees, open space, clean air, and the likelihood someone interesting might show up at the birdfeeder.

We can make several broad statements about the value of ecosystem goods and services and all of them relate to geography:

- The scarcer an ecological feature, the greater its value.
- The scarcer the substitutes for an ecological feature, the greater its value.
- The more abundant the complements to an ecological feature, the greater its value.
- The larger the population benefiting from an ecological feature, the greater its value.
- The larger the economic value protected or enhanced by the feature, the greater its value.

New York's Central Park makes this point clearly. It is one of the most valuable sources of ecosystem services in the world. Central Park isn't particularly desirable ecologically, but it is nevertheless valuable because so many people live near it and have so few substitutes within walking distance. Geography tells us about all of the factors noted above. We can map population densities, measure distances to similar parks, and easily detect the presence of other types of recreational open space and forms of access like roads.

The general proposition holds for most kinds of ecosystem services. The value of irrigation and drinking water quality depends on how many people depend on the water—which is a function of where they are in relation to the water. Flood damage avoidance services are more valuable the larger the value of lives, homes, and businesses protected from flooding. Species important to recreation (for anglers, hunters, birders, and the like) are more valuable when more people can enjoy them.

Placing a value on ecosystem goods and services also requires us to analyze the presence of substitutes for the good. The value of any good or service is higher the scarcer it is. How do you measure the scarcity of an ecosystem good? If recreation is the source of benefits, substitutes depend on travel times. What are walkable substitutes? Driveable substitutes? The value of irrigation water depends on the availability (and hence location) of alternative water sources. If wet-lands are plentiful in an area, then a given wetland may be less valuable as a source of flood pulse attenuation than it might be in a region in which it is the only such resource. In all of these cases, geography is necessary to evaluate the presence of scarcity and substitutes.

Finally, many ecosystem goods and services are valuable only if they are bundled with certain manmade assets. These assets are called "complements" because they complement the value of the ecosystem service. Recreational fishing and kayaking require docks or other forms of access. A beautiful vista yields social value when people have access to it. Access may require infrastructure—roads, trails, parks, housing. Note that these complements may themselves not be transportable. Again, neighborhood matters.

There are exceptions, in which geography is less important to valuation. For example, many of us value the existence of species and wild places *wherever they are*. When it comes to these kinds of ecosystem goods and services, location doesn't matter to our enjoyment, as long as the services exist somewhere. Another important clarification is that everything in nature is valuable if it contributes to the health of the overall system. Here, though, the value arises from the way nature produces services (the realm of the biophysical sciences). When it comes to the consumption of ecosystem goods and services, value tends to be determined by the social neighborhood.

Geographic Information as Technological Revolution

Geographic science will be challenging for both ecologists and economists of ecosystem services. The good news is that our technologies, data, and culture are becoming rapidly more map-focused. Armchair cartographers can already do amazing things with application platforms such as Google Earth. Government agencies and conservancies are making maps available that allow us to see both natural and social landscapes with remarkable detail. This technological revolution is having a cultural effect: maps are everywhere, changing the way we communicate and helping develop our spatial understanding of social and natural phenomena.

The growing deployment of geographic information systems is not without teething problems, however. This is particularly true when it comes to government creation and distribution of geographic information. The U.S. Census Bureau, for example, produces massive quantities of geospatial information on households and businesses. The integration of this information into widely shared, open-source software applications remains awkward, however. Private individuals are stepping in to help solve these problems, but much more could be done by government providers to aid the distribution of geographic information.

Nature is as important to our economy as are farms, factories, and multi-National corporations.

A larger worry is the lack of systematically and consistently tracked environmental information by our government trustees—a worry amply documented by the Government Accountability Office and other watchdog organizations. The greatest need facing us is to understand how we can protect and enhance ecosystem services and predict their loss. Geographic analysis of biophysical production functions is the key. But geographic analysis will rely on detailed ecological information tracked consistently over time. Unfortunately, agencies like the U.S. Environmental Protection Agency, NASA, the U.S. Geological Survey, and the Department of the Interior, among others, are given scant resources and authority to gather such information. Nature is as important to our economy as are farms, factories, and multi-national corporations. Geography is the key to understanding that economy.

The Agricultural Impact of Global Climate Change: How Can Developing-Country Farmers Cope?

NATHAN RUSSELL

No one understands better than farmers how weather affects people and their land, especially when it takes a turn for the worse. That's why farmers around the world talk and worry about the weather obsessively. But now, emerging weather patterns have a lot of other people worried, too, and their concerns are well-founded.

According to a recent report of the U.N. Intergovernmental Panel on Climate Change, the average temperature of Earth's surface, having already risen by 0.7 degrees Celsius in the last century, is expected to increase by an average of about 3 degrees Celsius over the next century, assuming greenhouse gas emissions continue to rise at current rates. Even the minimum predicted temperature increase, 1.4 degrees Celsius, will represent a profound change that is unprecedented in the last 10,000 years.

The scientific evidence behind these projections is unequivocal, leaving "no doubt as to the dangers mankind is facing," says Yvo de Boer, executive secretary of the U.N.'s Framework Convention on Climate Change. One of the gravest concerns facing the world, and especially developing countries, is agriculture's vulnerability to climate change.

The Risk Factors

Part of the danger to agriculture comes from extreme weather events, such as stronger storms, longer droughts and prolonged flooding events. Using computer-based simulation models, scientists predict these extreme events will occur with greater frequency, especially in the tropics. In addition, more subtle changes in rainfall patterns, together with rising temperatures, may shorten growing seasons in some areas, reducing crop productivity.

Never before have so many people been vulnerable to weather fluctuations, partly because never before has Earth been so densely populated, and partly because so many rural populations are poor. According to World Bank estimates, rural populations account for nearly 75 percent of the approximately 1.2 billion people who live in extreme poverty. Poverty, because it limits options, is a major reason that developing-country farmers are vulnerable to global climate change.

Another factor is the steady degradation during recent decades of the soil, water, forests and other plant resources on which rural livelihoods depend. This has resulted in large part from the intensification and expansion of agriculture in response to a growing demand from Earth's rapidly expanding human population for food, feed and fiber. Lacking more sustainable alternatives, farmers have often been obliged to adopt practices—such as continuously growing one crop and overusing chemical fertilizers and pesticides—that lead to biodiversity loss, soil erosion and water contamination.

Sub-Saharan Africa, the world's poorest region and one very dependent on agriculture, brings the problem into sharp focus. An estimated 500 million hectares of its agricultural land are already degraded, according to soil scientists. Moreover, 95 percent of the region's cropland is rainfed, according to the U.N. Food and Agriculture Organization, and rainfall patterns are already quite erratic.

With climatic challenges on the horizon, how farmers in Africa and elsewhere will keep pace with the demand for food remains to be seen. But their only hope of creating better livelihoods for themselves is through income growth and effective stewardship of natural resources.

Mapping the Menace

The potentially dire consequences of global climate change, to paraphrase literary figure Samuel Johnson, are wonderfully concentrating the minds of agricultural scientists, development professionals and policymakers around the world.

Among them are the approximately 8,000 scientists and staff of the 15 centers supported by the Consultative Group on International Agricultural Research, or CGIAR. Its work represents the largest public investment in international agricultural research. The CGIAR centers and their partners in governmental and nongovernmental organizations have been working for years to help farmers cope with the effects of variable and severe weather.

CGIAR scientists have significantly advanced the understanding of what specific consequences rural populations will face in particular places as a result of climate change. Several years ago, researchers in Africa and Latin America, using weather and crop simulation models, mapped the impact on both regions' maize production. Models predict a 10 percent decline in maize productivity by 2055. Moreover, the results reveal significant variation in the effects of climate change from one place to another, indicating where devastating maize crop losses could be expected.

Planning for Food after "Doomsday"

Last month, construction began on what's being called the Noah's Ark of seeds, or the Doomsday Vault. The Svalbard International Seed Vault is a "fail-safe" vault that will hold seed samples of most food crops from most countries. It is designed to protect the essential agricultural seeds from damage or loss due to climate change, war or any other disaster, according to the U.N. Food and Agricultural Organization's (FAO) Global Crop Diversity Trust, which is co-funding the project with the Norwegian government.

The vault is being carved 120 meters deep into an island not far from the North Pole. Researchers chose the site partly because the ground is perpetually frozen (permafrost), thus providing natural refrigeration for the seeds, and partly because at 130 meters above sea level, the site is high enough not to be affected by even the most drastic rising seas should all polar ice melt. Deep inside the island, the seeds will be kept at a cool 10 to 20 degrees Celsius below zero, so even if air temperatures rise significantly and much permafrost melts, the site should be secure, according to researchers involved in the project.

About 1,400 seed banks, or genebanks as they are often called, exist worldwide, ranging in size from one type of seed to more than 464,000 different samples in the U.S. genebank, according to FAO. Exactly how many and what type of seeds will be in the Svalbard vault is ultimately up to each country that sends in its seeds. The vault is capable of holding up to 3 million seed samples, which would make it the most comprehensive in the world.

Seed banks are important for preserving biodiversity and genetic variation, researchers say. Different types of crops have different genetic characteristics, which may make one variety resistant to disease and another resistant to drought. For example, more than 100,000 varieties of rice exist, each with varying genetic traits. Still, a large number of crop varieties are thought to have already been lost through history, according to FAO. For example, of the 7,100 named apple varieties grown in the United States in the 1800s, more than 6,800 no longer exist.

"Every day that passes we lose crop biodiversity," said Cary Fowler, executive director of the Global Crop Diversity Trust in a Feb. 9 press release. "We must conserve the seeds that will allow agriculture to adapt to challenges such as climate change and crop disease," he said.

The vault, which is scheduled to be finished in September, is part of a larger global strategy being implemented by the Diversity Trust to protect collections of crop genetic diversity from the ravages of natural disasters, rising temperatures, rising sea levels, floods, drought, and disease, as well as from human-induced issues such as wars, civil strife and accidents.

—Megan Sever

Researchers have also mapped the impacts of climate change on wild species related to three food crops: cowpea, peanut and potato. If the climate changes as researchers expect, 22 out of 51 wild species of peanut, for example, will most likely become extinct by 2050, and the area covered by these species will shrink to 10 percent of its present size.

In addition to foreseeing the fate of plants on which people depend, CGIAR scientists are mapping the vulnerability of agricultural systems, based on a variety of biophysical and social factors. A recent study carried out in sub-Saharan Africa, for example, identified "hotspots" of vulnerability to the combined menace of climate change and poverty.

The maps produced by such studies provide an early warning as to which plant species, agricultural systems and rural communities are at greatest risk in the long term.

Climate-Resilient Food

If farming communities are to successfully adapt to climate change, they will need crop varieties and livestock breeds possessing greater tolerance to stresses such as drought and heat. Research has significantly advanced in developing hardier crop varieties through conventional breeding methods, and the application of tools from molecular biology is speeding the process. One challenge is to make more extensive use of crops' wild relatives. These contain genes for traits such as drought tolerance, which could prove useful for adapting crops to harsher conditions. "Climate-resilient" varieties resulting from crop improvement programs have already reached farmers' fields, and more are in the pipeline.

Maize breeders working in sub-Saharan Africa, for example, have developed more than 50 new varieties that are tolerant to drought, and these are already being grown on at least a million hectares. Part of the secret behind this triumph is a novel breeding method, in which hundreds of small farmers take part in testing new varieties under harsh growing conditions. Varieties selected on the basis of field testing at nearly 150 stress-prone sites in eastern and southern Africa are yielding 20 percent more, on average, than the ones they replace, and the best new varieties are doubling yields.

On the other side of the world, rice researchers have found genetic relief from floods. They have identified a rice gene that allows plants to survive when completely submerged for up to two weeks, while most rice survives underwater for only three days. The "waterproofing" trait has been transferred into a popular rice variety in Bangladesh, and is helping farmers reduce crop losses from flooding.

Among the world's most naturally hardy food crops are barley, millet and sorghum, which are widely grown in dry climates. Barley breeders in Syria have demonstrated how drought tolerance in this crop can be markedly improved through a method involving farmer participation, and the approach is now being applied in many other countries of the Middle East and North Africa. Using tools from molecular biology, researchers have isolated and are employing the so-called stay-green trait in millet and sorghum to bolster their drought tolerance.

Researchers are also looking to bolster livestock breeds and their food sources, and are currently introducing data on the distribution of African livestock breeds into computer-based geographical information systems, or GIS. When overlaid with climate change and ecological data, this information will help select breeds for environments where drought is becoming more prevalent. Additionally, to secure food sources for livestock, researchers are selecting and promoting drought-tolerant grass and legume species.

Apart from causing direct damage to crops and livestock, higher temperatures and related changes will expose them to further depredations of diseases and pests, which already take a heavy toll on developing-country agriculture. To anticipate and prepare for this problem, CGIAR scientists are examining the likely effects of climate change on major biotic stresses in agriculture, including certain human and animal diseases such as malaria and trypanosomiasis.

Changing the System

The performance of crops and livestock under stress obviously depends not just on their inherent genetic capacity but on the whole system in which they are produced. For that reason, any serious effort to increase the resilience of developing-country agriculture in the face of climate change must involve more prudent management of crops, animals and the natural resources that sustain their production while providing other vital services for people and the environment. Water is an especially critical resource, and its management is closely intertwined with that of soil and biodiversity, including forests.

Research in improving the management of natural resources in developing countries is generating new knowledge and tools that are highly pertinent to the task of helping farmers cope with changing climatic conditions. For example, a recently completed "comprehensive assessment" of water management over the past 50 years points to a wide range of proven technologies, such as water harvesting and simple irrigation systems, as well as policy options for increasing water productivity in both irrigated and rainfed agricultural systems, including livestock and fisheries.

Similarly, international networks of tropical soil scientists—such as the African Network for Soil Biology and Fertility—have devised integrated approaches to improving soil fertility that combine targeted application of inorganic nitrogen and phosphorus fertilizers with the use of livestock manure and other locally available organic sources of nutrients.

One such nutrient source consists of various "agroforestry" species—especially of the genera *Gliricidia, Sesbania* and *Tephrosia*—that thousands of farmers in southern Africa have adopted to raise soil fertility on fallow land. Agroforestry is the practice of integrating "working" trees into agricultural landscapes as natural vegetation is cleared. In addition to maintaining soil health through nitrogen fixation and use of cuttings as fertilizer and mulch, these trees provide useful products, such as animal fodder, fruit, timber, fuel, medicines and resins.

CGIAR forestry research has produced a rich collection of knowledge and tools that provide conservationists and others with better means of monitoring forest management and certifying whether it is sustainable in particular cases. Of the total forest area certified so far, more than 80 percent (some 37 million hectares) has been certified by companies that acknowledge using the products of this research. This has resulted in better forest management, contributing to more sustainable livelihoods for forest dwellers.

With the knowledge and technology already available, large improvements can be made in natural resource management. These are already imperative and will become more so in the coming decades both to help mitigate climate change—that is, reduce greenhouse gas emissions through increased capture of carbon in trees, agroforestry species and crop residues left on the ground—and to better enable farmers to cope with climate change. Without help, developing-country farmers will stand little chance of success.

Markets and Models

As compelling as this logic may be, fostering widespread use of knowledge and practices aimed at protecting natural resources is not easy, though it is being accomplished. Three conditions must be met to accelerate the process: strong incentives, able institutions and supportive policies.

Researchers are exploring avenues that enable the rural poor to afford investing in improved management of natural resources. A central challenge is to link rural communities more strongly with markets for higher value products and services, such as horticultural crops, tropical fruits, livestock products, ecotourism and a variety of environmental services.

One promising option involves the new and rapidly growing world market for certified reduction of carbon emissions. Under the Clean Development Mechanism established by the Kyoto Protocol, countries that do not meet their agreed targets in emission reductions can buy the service from other countries. Carbon funds have been established for this purpose, but so far they have traded mainly with sectors such as energy and transportation in developed countries.

Some schemes, however, such as the BioCarbon Fund, have been set up that cater specifically to agricultural and forestry projects. But limits on payments to such projects and the complexity of the procedures pose significant barriers to participation, especially for small farmers.

Exciting work is under way on various fronts to reduce the obstacles to carbon trading. For example, CGIAR agroforestry researchers have devised and are applying a new technique in eastern Africa that assesses soil conditions, including carbon stocks, with a high degree of accuracy. Involving the use of satellite imagery and infrared spectroscopy, the technique is much cheaper than on-the-ground verification.

Scientists are also developing simulation models that will enable them to provide comprehensive assessments of the many factors affecting food security, poverty and the environment as influenced by climate change. Such information is critical for enabling policymakers and others to define a vision of the way forward toward sustainable development and to design measures that will help realize that vision.

This vision must include new institutional arrangements at the local, national and international levels that will facilitate the participation of small farmers and other land users in carbon trading.

Let's Be Civilized

Developing-country agriculture is one of the central arenas in which the threat posed by climate change must be confronted. The efforts under way so far provide part of the basis for action, but they must be more sharply focused and better coordinated.

Much depends on the success of collaboration. Agriculture, after all, still forms the basis of "civilization," a concept that has less to do with material and cultural progress, according to world-renowned historian Felipe Fernández-Armesto, than with how people shape and adapt to diverse environments in meeting their food and other needs.

Global climate change poses a new challenge to humanity's skill at maintaining viable livelihoods under highly diverse and variable climatic and environmental conditions. We might even think of it as the ultimate test of just how civilized we can be.

NATHAN RUSSELL is a senior communications specialist with the Secretariat of the Consultative Group on International Agricultural Research (CGIAR).

When Diversity Vanishes

Don Monroe

Complex systems, from ecologies to economies, do interesting and unexpected things. Much of this rich behavior can be traced to the networks through which the underlying "agents" affect each other. Often, however, diversity of the agents themselves is essential. If they act too similarly, the entire system can cease to function. At the annual symposium of the Santa Fe Institute Business Network, November 1–3, 2007, an array of experts explored this "diversity collapse" in contexts ranging from ecology and the food we eat, to finance and organizational structure.

Ecological Collapse

The most dramatic diversity collapses are mass extinctions, which have wiped out much of life five times in Earth's history. Doug Erwin, of the National Museum of Natural History and SFI, said that one of these, the end-Permian extinction, wiped out "90 to 95 percent of everything in the oceans, about 70 percent of everything on land, and by all accounts was about the best thing that ever happened to life on Earth." The extinctions made room for later innovation—but not right away. "Eventually the diversity got bigger than before, but it took four million years to even get started."

In contrast to some observers, Erwin does not believe that we are entering a "sixth wave" of extinction. "At least if we're lucky, we're not," he said. Nonetheless, "the crisis is real." Erwin emphasized that there are many types of diversity, which do not have the same impact. For example, individual species on different branches of the tree of life forms can become extinct without substantial effect, but losing the same number of species on a single branch could eliminate that entire branch.

Global extinction reflects the combined changes in smaller, individual ecosystems around the world. Andrew Dobson of Princeton University described what he called "probably the best-studied" ecosystem: the Serengeti National Park in Tanzania. Established in 1951, this park and the surrounding areas provide "a natural example of what happens when we perturb an ecosystem," he observed. Outside the park, the ecology changes dramatically because of farming and grazing. The difference is most notable at the highest trophic levels in the food chain, Dobson observed.

Dobson contributed to the Millennium Ecosystem Assessment, which framed the contributions of healthy ecosystems, at least in part, in terms of the economic "services" they provide to people. Almost half of the value, Dobson said, comes from the most basic level, including bacteria, and another one third from plants. These lower levels also tend to be more resilient. Higher trophic levels, including grazers and predators, are more visible but provide less value, he said. They are also more sensitive to changes, so "monitoring these brittle species gives an early warning" of damage.

"We have a scarily short time scale to understand how ecosystems collapse," Dobson commented. "Most large natural ecosystems will be destroyed in the next 30 to 50 years. The quality of human life on this planet is dependent on the economic services supplied by those webs."

Historically, said Mercedes Pascual of the University of Michigan and SFI, ecologists viewed complexity in food webs as an essential feature of healthy ecosystems that helps them to resist disruption. In contrast, monocultures, such as the endless fields of U.S. Midwestern corn, can succumb to a single pest.

The important work of Robert May in the 1970s, however, showed that complexity actually reduces stability in some mathematical models. Ever since, Pascual said, ecologists have tried to understand "how more realistic structures lead to higher stability."

Instead of studying small perturbations as May did, Stefano Allesina, of NCEAS, and Pascual looked at major shifts such as the disappearance of prey causing a predator to become extinct. They also used food webs taken directly from ecological studies instead of mathematically generated networks. They then determined which pathways are functional and which are redundant, from the perspective of secondary extinction, and found that in real webs about 90 percent of connections are functional, independent of the size of the ecosystem.

As a result, Pascual observed, "Even when secondary extinctions are not observed, the loss of species makes ecosystems more fragile to further extinctions." There may be little warning of an approaching "tipping point," in which the entire ecosystem collapses.

Pascual suggested that work by SFI External Professor Ricard Solé may clarify the dynamics of interacting populations contributing to extinctions in these ecological networks. In his model, as species continually become extinct and new ones immigrate into a region, the food web forms a self-organized state, with many of the features observed in real webs.

Although the system as a whole seems static, individual species are not. "Individual populations . . . are all going up and down like crazy," Pascual said. These fluctuations at one level may even enhance the stability at a higher level. The extinction of individual species may therefore be a misleading measure of the loss of diversity.

The question of the best level for gauging diversity also arose in work by Katia Koelle (Penn State), Sarah Cobey (University of Michigan), Bryan Grenfell (Penn State), and Pascual on the evolution of flu. Genes evolve continuously, but often with no effect on the "phenotype": the surface proteins that determine immune response. The researchers modeled genetic evolution coupled with the prevalence in the human population of immunity to particular variants. In this model, viruses multiply rapidly whenever they take on a new phenotype, quickly crowding out other variants. "This pattern of boom and bust is explained by an interaction of genetic drift and selection, and not exclusively one or the other," Pascual said.

Managed Ecosystems

If the ecologists are right, natural ecosystems, which have evolved complex webs of interactions, are ideally "managed" by leaving them alone—when possible. In stark contrast, agricultural crop management resembles control theory—an engineering tool developed for much simpler systems—and features simplified ecosystems that depend heavily on external inputs. Much of the corn in the U.S. is grown with petroleum-derived pesticides and fertilizer, and fed to cattle whose excrement then becomes toxic waste rather than nutrition for plants.

An alternative was described by Joel Salatin, who recovered marginal land in Virginia by using cow manure to fertilize the grasses that are the natural food for cattle. But Salatin complained that the many regulations aimed at industrial-scale production present formidable barriers to small farms like his.

In spite of such efforts, local production is likely to be an anomaly in an industrial food system that prizes cheap and abundant food. But Cary Fowler of the Global Crop Diversity Trust asserted that this system depends on underappreciated diversity of plant varieties.

Even as agriculture has focused on fewer species over the past 12,000 to 15,000 years, Fowler said, "diversity in some sense was increasing," as farmers in different regions selected variants with different traits. "There are about 120,000 different varieties of rice, each as distinct one from the other as a Great Dane from a Chihuahua," he commented.

Industrial agriculture, Fowler said, threatens this variation within species, although scientists still cannot agree on how to measure it. In response to new challenges, he wondered, "can we continue to develop our agriculture without diversity?" His answer: "Obviously we can't."

The expected global warming in coming decades makes these issues especially urgent. "My guess is we are ill-prepared for this kind of change," Fowler said. "But if we are prepared, it will be because of the gene banks and the diversity they contain."

> **"My guess is we are ill-prepared for this kind of change," Fowler said. "But if we are prepared, it will be because of the gene banks and the diversity they contain."**

Numerous gene banks have been storing seeds for crops and other plants around the world. Unfortunately, Fowler said, many of them are poorly funded and maintained. He stressed that for a modest cost—an endowment of some $250 million—"we can conserve the gene pool of our major crops in perpetuity." As a start in this direction, Fowler's organization is funding a facility above the Arctic Circle to provide a global seed repository, sometimes called the "doomsday vault."

Valuing Diversity

No such vault exists to preserve human culture. "The forces of homogenization are rampant," said Suzanne Romaine of the University of Oxford. She described the rapid extinction of languages, as large ones like Mandarin, Spanish, and English spread at the expense of smaller ones.

As a linguist, Romaine values languages as data for her own work. But she sees linguistic extinction as part of a larger problem. "It's not just languages that are at stake, but forms of knowledge," she said. "They can't be separated from people, their identities, their cultural heritage, their well-being and their rights." She also stressed that language diversity and biodiversity often disappear together.

In a similar way, shared communication drives a homogenization in computer systems, said Gabriela Barrantes, of the University of Costa Rica and SFI. The dominance of Microsoft in personal computer software is only the most visible example of this diversity collapse, she said.

This uniform computing environment is sensitive to threats, just as monocultures are vulnerable to agricultural pests. Barrantes and Stephanie Forrest, of the University of New Mexico and SFI, have been exploring how artificial variability in computer systems can slow the spread of malware. In any successful scheme, she stressed, computers must still interoperate with similar performance and cost.

Diversity can be introduced at many levels, Barrantes said. For example, the well-known "buffer overflow" attacks rely on long data spilling into areas of memory intended for programs. Varying the locations of these segments can often thwart the spread of infection between different machines. This kind of artificial diversification is "currently being used in two major operating systems," Barrantes said.

Diversity collapse in computer systems is probably well ahead of that in ecosystems, Forrest suggested. But she noted that "adding diversity back in is much easier than it would be in the natural world."

The dominance of Google, eBay, and others shows that "online niches are often winner-take all," notes Virgil Griffith of the California Institute of Technology and SFI. But Griffith claimed that "promiscuous interoperability" can promote diversity by allowing people to use data for new purposes. "When all-powerful monocultures make data available, diversity flourishes," Griffith claimed. "The users diversify the monoculture, not the other way around."

Diverse Perspectives

In finance, diversification reduces risk by spreading money among assets that respond differently during market moves. But during the 1998 international financial crisis, Long-Term Capital Management suffered enormous losses when its ostensibly diversified investments began to react similarly as other investors desperately sold the same assets. "Diversity collapses are really the source of inefficiencies in markets," asserted Michael Mauboussin, Chief Investment Strategist at Legg Mason Capital Management and an SFI trustee.

Mauboussin reviewed three theories for how markets become efficient, meaning that prices reflect value. The first, in which all investors behave rationally, is unrealistic. A second explanation, which requires only that some investors exploit—and thereby remove—arbitrage opportunities, "has failed us in critical junctures," he asserted.

Mauboussin contrasted these models with a view of "markets as a complex adaptive system, where prices essentially emerge from the interaction of many agents." In this view, also called "the wisdom of crowds," three conditions assure efficiency: diversity among investors, an aggregation mechanism, and financial incentives. Of these assumptions, he said, "the most likely to be violated is diversity," which may decline imperceptibly until it suddenly collapses.

Scott Page of the University of Michigan and SFI, has compiled many ways that diverse groups outperform individuals. In expert judgment, for example, as in diversified portfolios, the average judgment of a group is always better. "This is not a feel-good statement, this is a mathematical theorem," Page said. For problem solving, having multiple strategies can help a group evade roadblocks that hamper any one approach.

"In human systems," Page said, "the thing that really works against cognitive diversity is selection." The increasingly global marketplace of ideas selects the current "best practices" at the expense of other approaches. "If the world is flat, we cannot count on the right amount of diversity existing," Page said.

Page warned that merely recognizing the advantages of cognitive diversity might not be enough to preserve it. "Diversity's benefits may be public goods that are not in any one's interest to maintain."

One way to maintain diversity is through time-varying selective pressures that prevent one idea from dominating. However, Page showed a simple model in which such churn did not prevent uniformity. He suggested that maintaining diversity also requires diverse selective processes or richer networks, so that the criteria for picking winners varies.

The broad range of speakers at this symposium shows that the Santa Fe Institute is in no danger of diversity collapse, although similar principles apply in very different fields. Still, in the world outside, increasing interconnectedness seriously threatens diversity in both human organizations and ecosystems.

DON MONROE (www.donmonroe.info) writes on biology, physics, and technology from Berkeley Heights, New Jersey. Prior to 2003, he spent 18 years in basic and applied research at Bell Laboratories.

Classic Map Revisited
The Growth of Megalopolis

Browning's classic 1974 map of Megalopolis, covering the growth of Megalopolis from 1950 to 1970, is updated through 2000. The color map depicts the extent and expansion of Megalopolis for three time periods, 1950–1970, 1970–1990, and 1990–2000. Discussion relates the growth of Megalopolis to social and economic forces influencing urbanization in the United States in the latter half of the twentieth century.

RICHARD MORRILL

For urban and population geographers, a wonderful and valuable map was produced by Clyde Browning and was published in 1974 in the University of North Carolina's *Studies in Geography*. Browning's article, "Population and Urbanized Area Growth in Megalopolis, 1950–1970," was both a quality representation of Megalopolis and an updating of its expansion through 1970. The term *megalopolis* had been coined by Jean Gottmann, recognizing the string of urbanized areas extending from Boston to Washington, D.C., as the "main street of America" (Gottmann 1961).

Megalopolis now extends from Fredericksburg, south of Washington, D.C., to Portsmouth and Dover-Rochester, into southern Maine. My updated map depicts the further expansion of the nation's largest conurbation, whose constituent parts housed 24.5 million people in 1950 and 42.4 million in 2000.

This brief article has two purposes, first to depict and appreciate change in the size and spatial pattern of Megalopolis, inspired by the classic Browning map, and second, to review the changing forces that have shaped this remarkable product of human settlement.

Megalopolis and Its Mapping

Many academics have attempted to coin terms for their phenomenon of study, but few such terms have been successful. Gottmann's term *megalopolis* to refer to a string of closely interconnected metropolises was logical and inspired and has become part of the language (Kahn and Wiener 1967). *Megalopolis: The Urbanized Northeastern Seaboard of the United States* (Gottmann 1961) was a massive undertaking (more than 800 pages), of detailed scholarship and amazing insight—a tracing of the evolution of the "main street" of then US1 to the interconnected promise of I-95.

Part 1 of *Megalopolis* argues the dynamic role of the core cities from Boston to Washington, D.C., in the economic and cultural development and control of the nation, the "economic hinge" of innovation, including suburbs as early as 1850. Part 2 concentrates on the structure of population and land use, especially in the suburban fringe, noting the long-standing but now faster growing penetration of urban uses into the country (i.e., sprawl), again long before other parts of the country noticed. Perhaps there was a higher expectation of the survival of close-in agriculture than has proven possible. The beginnings of urban decay and of renewal are treated, with a plea for rehabilitation instead of renewal that was finally successful in the 1980s and 1990s.

Part 3 details patterns of economic structure and change. The chapter on the white-collar revolution, outlining the restructuring to higher level activities, is probably the most important and prophetic analysis in the book, already predicting in 1960 the basic remaking of American society, with the Boston to Washington, D.C. Megalopolis leading the way. Part 4, "Neighbors in Megalopolis," recognizes the diversity and segregation of the population along ethnic, racial, religious, and class lines; the high level of inequality that characterizes creative cities; and, finally, the difficulty of coordinating planning across utter jurisdictional complexity.

Gottmann later compared Megalopolis to other world megalopolitan systems (Gottmann 1976), and still later revisited Megalopolis in *Megalopolis Revisited: 25 Years Later* (Gottmann 1987). He was able to see the validation of his restructuring prediction and the incubator role of Megalopolis, and especially of New York. Yet he notes as well the pace of deconcentration within Megalopolis.

Browning's 1974 map was quite a large and detailed representation of Megalopolis, tracing its expansion to 1960 and to 1970. It is not practical to attempt to reproduce the original here. The monograph text presents a thorough empirical and theoretical discussion of the magnitude and nature of change. Browning provides an overview of urbanized areas and of Megalopolis, and a statistical and graphic summary of the change from 1950 to 1970, noting that not only had most cores not coalesced, but that the metropolitan region defined by Gottmann was still less than 20 percent urban territory. This is followed by a tight review of classic urban growth theory, based on work of Mayer (1969), including reproduction of an amazing map of "The Region's Growth" (Regional Plan Association 1967). The monograph then provides case studies of Boston (by Conzen), Rhode Island (by Higbee and Higbee), New York (by Carey), Philadelphia (by Muller), and Washington (by Brodsky), in which a common theme is the discontiguity of suburban growth and the role of physical and institutional barriers.

The Updated Map: Change from 1950 through 2000

The original 1974 map, covering 1950 through 1970, was based on urbanized area delineation, which in turn relied on corporate boundaries and enumeration districts, and rarely took into account over-bounded cities. For this reason, the updated map, presented here as Figure 1, uses the census tract as the basic unit, rather than the far more precise delineation possible with block groups in 1990 and blocks in 2000. As a result the map does suggest a greater areal coverage than do the 2000 maps of urbanized areas, but it is also obvious that all this territory and more is functionally part of the intense commuter labor market of the metropolitan centers, and gives a realistic sense of metropolitan dominance. The 1950 urban cores represent the end of the era of central city dominance and dense urban settlement. The 1950 to 1970 change resulted from the postwar housing and suburban boom. The 1970–2000 change starkly captures the impact of

metropolitan growth in the context of rising affluence, and the massive decentralizing role of the Interstate Highway System (Frey and Speare 1988). Megalopolis may not be as dynamic as the rising south or the burgeoning west, but it is still number one—the largest and most important metropolitan region.

The updated map for 2000 adds areas that became urban by 1980, 1990, and 2000, not only to places that were already urbanized areas in 1950–1970, but places that became urbanized areas over the thirty years that followed. For maximum comparability, the delimitation of areas for all the censuses 1950 to 2000 is based on the 2000 census tracts. From Table 1, we can see that the *population* of Megalopolis has not quite doubled, but the total *area* has quadrupled, and mean densities have fallen from 7,315 to 3,155 persons per square mile.

Consider the first (1950) and last (2000) stages illustrated in the map. In 1950 Megalopolis was actually a "string of pearls" composed of Washington, Baltimore, Wilmington, Philadelphia, Trenton, New York, Bridgeport-New Haven, Hartford, Springfield, Providence, Worcester and Boston, Lowell, and Lawrence—all distinct places, separated by some rural territory. These were the core urban places that had arisen in the colonial period and they had exhibited an extraordinary linearity, based partly on physical character (the head of navigation at the fall line) and partly on the situation, sea or river ports and early industrial centers, convenient for trade with Europe (Dunn 1983).

By 1970, Wilmington-Philadelphia-Trenton had merged, as had Boston-Lowell-Lawrence, but, perhaps surprising to many, no others had merged although there had been very significant suburbanization, especially around New York and

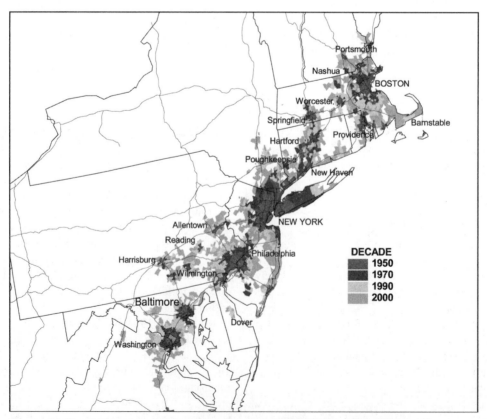

Figure 1 Population and urbanized area growth in Megalopolis, 1950–2000.

Table 1   Population in Megalopolitan Urbanized Areas (millions)

Year	Population	Area[a]	Density	Year	Population	Area[a]	Density
1950	24.5	3283	7315	1980	34.4	8390	4100
1960	29.4	5348	5285	1990	36.6	10185	3590
1970	34.0	7006	4768	2000	42.4	13490	3155

[a]Area in square miles.

Washington, D.C. New urbanized areas included Vineland, Danbury, Fitchburg, and Nashua.

By 2000 a continuous urban settlement structure for Megalopolis had almost been realized, with a smaller Washington-Baltimore-Aberdeen to the south, a giant Wilmington to Springfield and Norwich in the center (with links to formerly independent places like Atlantic City, Allentown, Lancaster, York, Harrisburg, and Poughkeepsie), and a northern area from Providence and Barnstable through Boston to Manchester, Portsmouth, and Dover-Rochester. New outlying urbanized areas, not yet quite connected, include Fredericksburg, Dover, Wildwood, Frederick, and Kingston. The map graphically captures the massive urban diffusion from early cores, the gradual coalescence of these expanding cores, and the rise of and reaching out to satellite places (Dorgan and Kasarda 1988).

Methodology: Reporting Units and Classification Criteria

For all six censuses the basic units for the delimitation of Megalopolis were the constituent census-defined urbanized areas—that is, urban agglomerations with populations greater than 50,000, and with urban territory consistently defined as contiguous areas with densities of more than 1,000 persons per square mile. But because the available building blocks and the criteria for delimitation of urban from rural varied somewhat over the years, some standardization and consistency was achieved by using the latest 2000 census tract geography as constant units for analysis over the decades.

The extent of megalopolis was delimited by superimposing the 2000 census tracts over the urbanized area extent for each decade. For 1990 and 2000, the nation was blocked, and a more precise delimitation of urbanized areas was possible. But because such detail did not exist before 1990, or for the original Browning map, census tracts proved the most effective unit for comparative analysis over the entire period. However, use of census tracts rather than block groups or blocks for 1990 and 2000 does result in the inclusion of some rural territory and in the exclusion of some urban territory in the final map, and a generalized rather than a spidery urban edge. This smoothing means a slightly more extensive coverage for 2000 than the block-based urbanized area mapping from the Census. On the other hand any rural territory in the included tracts is obviously a functional part of the interconnected labor markets and their metropolitan centers and gives an accurate representation of metropolitan dominance. If the presence of high levels of commuting to metropolitan centers were used to depict the extent of Megalopolis, as indeed Gottman invoked in the original book, the area would be more extensive than shown in the updated map, but not dramatically so (Morrill, Cromartie, and Hart 1999).

Forces for Change in Megalopolis, 1950–2000

The second half of the twentieth century was an era of continuing metropolitan expansion in the United States, exemplified well by the changing map of Megalopolis (Muller 1981). It is useful to summarize briefly the forces that produced these patterns of settlement change, even though this has been the subject of countless studies in several disciplines (e.g., see Johnston 1982; Castells 1989; Knox 1993; Orfield 2002).

In the case of Megalopolis, the underlying set of cores has been established for a century or more. The settlement processes that have dominated in the past fifty years include (1) sheer economic and demographic growth, (2) physical decentralization in the form of suburbanization, (3) extension of metropolitan commuting fields and the physical coalescence of formerly physically separate areas, (4) rise or restructuring of and reaching out to formerly distant satellites, and (5) restructuring and revitalization of high-level metropolitan cores. The first four are graphically represented on the map.

The fifty years may be usefully divided into three periods. The first, 1950–1970, was characterized by rapid growth and even more rapid suburbanization, and was the period corresponding to the original Browning (1974) map of Megalopolis. The second, 1970–1990, was one of some inner metropolitan decline and racial conflict, but continuing suburbanization and the rise of "edge cities." The third, since 1990, saw metropolitan core resurgence and gentrification, inner suburban maturing, and far-suburban and exurban and satellite city growth. These divisions are perhaps not perfect with respect to societal trends (1950–1965, 1965–1985, and 1985–2000 would have been preferred), but they proved best, given the need for using decennial census data, and, at least for Megalopolis, correspond to periods of faster, then slower, and then faster growth (Table 1).

1950–1970: The Rise of the Suburbs

Suburban growth was pervasive over most of what is now Megalopolis, fueled by the high fertility and natural increase of the baby boom as the nation reacted to losses from World War II, and by very large domestic rural-to-urban migration. Almost all industrial sectors and types of cities grew as part of postwar recovery, even the older industrial sectors and cities.

Metropolitan growth reflected the dominance of both increasing returns to scale and to agglomeration and the proliferation of new products and services. This growth was spatially expansive, via burgeoning new suburbs, mainly because of sheer population growth and preferences of the larger baby boom families, but, as has been endlessly repeated by urbanists, it was abetted by government housing policies and institutions (Federal Housing Administration, the GI Bill) and planning policies for separation of uses (and of races), and was encouraged by the first fifteen years of the Interstate Highway System. The period was also one of large-scale in-migration of blacks fleeing the more discriminatory South, which in turn precipitated large-scale white flight to the suburbs, notably around Washington, Baltimore, Philadelphia, and New York. Substantial suburbanization of industry and of shopping began to follow the suburban migration. The attractive pull of suburbs, both for families and jobs, dominated throughout the baby boom period until 1965; then by the late 1960s the partly perceived and partly real problems of inner city decline and disinvestment became very strong motivators for suburbanization.

1970–1990

The population in Megalopolis grew more slowly during this period, especially from 1970 to 1985, than in the preceding or the following periods, despite continuing suburban growth, because of often-declining absolute populations and, often, declining employment in the dense, older central city cores (Berry 1976). Indeed the 1970s were rare years of more-rapid nonmetropolitan than large metropolitan growth nationally, as the giant cities were beset by racial tension, large-scale white flight to the suburbs, and the decline of traditional industries, including manufacturing and transportation (Noyelle and Stanback 1984; Harvey 1989). Fertility and natural increase fell as the baby boom was replaced by the babybust. Megalopolis fared better than more industrial areas like Buffalo or Pittsburgh or Cleveland, but growth and prospects seemed dim in comparison to metropolitan growth in the west and south. Would the torch of "main street, USA" pass to the fast-growing Sunbelt cities? (Stanback 1991)

But Megalopolis did continue to grow in area—from 7,000 to more than 10,000 square miles, up 45 percent, even as population growth was a mere 7 percent as suburbanization continued and densities fell from 4,768 to 3,590 persons per square mile. Completion of the Interstate Highway System enabled and encouraged suburban growth, including large industrial and office parks oriented to external markets, and the shift from rail to truck long-distance transport. Suburban downtowns, termed "edge cities" (Garreau 1991), arose to challenge central city dominance (e.g., Tyson's Corner in Washington D.C.'s Virginia suburbs). Yet it proved premature to write off the old centers. Especially after 1980, the cities fought back, not by the unsuccessful urban renewal of the earlier period, but by deliberate investment in attracting higher class people and jobs through new sports arenas and arts complexes and the subsidization of high-rise office tower development—ironically aided by the same interstate highways that encouraged suburbanization of other branches of the economy. Perhaps this overstates the role

of the core cities. Alternatively, the urban cores were the places of highest metropolitan accessibility and existing infrastructure, and office and residential developers accurately perceived the long-term returns to reinvestment.

1990–2000

Even before 1990, much of Megalopolis experienced a revitalization and resurgence of growth in this latter period, with a hefty growth of 12 percent in the 1990s alone. Gottmann himself had already outlined the dimensions of this new urbanism in his book, *Megalopolis Revisited: 25 Years Later* (Gottmann 1987). A large literature on the contemporary city provides provocative and contradictory reading (e.g., Sassen 1991; Soja 2000; Wheeler, Aoyama, and Warf 2000; Batty 2001; Scott 2001).

The larger downtowns and nearby historic areas were gentrified as middle and upper class households reclaimed parts of the core (Smith 1986.) Economic restructuring, as presaged by Gottmann, massively increased service employment, as business services and finance demonstrated a preference for central high-rise venues. Core populations rose, in part by attracting young, later- or not-marrying professionals and empty-nesters (Florida's "creative class"; Florida 2002) and in part from a resurgent large-scale immigration, especially in the 1990s, from Asia, the Caribbean, and Eastern Europe. Economic restructuring and gentrification led to a much greater degree of social and economic inequality, and rising costs of core areas led to some displacement of the poor and of racial minorities to the older inner-suburban zones, which often suffered relative decline.

But growth was vibrant in the ever more distant suburban fringe as well, greatly exceeding in absolute population and jobs the revitalization of the cores, with continuing industrial, commercial, and residential expansion. Much of the growth could be termed as low-density exurban sprawl, but where "smart growth" urban planning came into vogue, some of the growth concentrated in older, formerly independent satellite towns and cities, now incorporated into the Megalopolitan web (Peirce 1993). The revitalized cores dominated selected service and finance sectors, and the far suburbs continued to be most attractive for wholesale and retail, transportation, manufacturing, and less-professional service activities. Despite core revitalization, mean densities continue to fall to less than half what they had been in 1950.

Finally, some of the growth in these far suburbs or satellite cities was fueled by families seeking affordable single family housing and suburban schools, people driven out by high housing costs and perceived inner city social and school problems.

Conclusion

It is reasonable to conclude, with Gottmann, that Megalopolis remains the Main street of America, despite the much faster rate and amount of growth in the metropolitan South and West. California may well be the trend setter of the nation in many ways, but Megalopolis remains the control center of our information economy and the innovator of urban settlement change, and has proven remarkably adaptable in maintaining its preeminence. The area defined as Megalopolis for the updated map housed

42,400,000 people in 2000. The exurban area surrounding Megalopolis, with high levels of commuting to megalopolitan jobs, housed at least eight million more. This amazing conurbation remains the most spectacular and powerful settlement complex and human imprint on the landscape.

References

Batty, M. 2001. Polynculeated urban landscapes. *Urban Studies* 38:635–55.

Berry, B. J. L., ed. 1976. *Urbanization and counter-urbanization.* Beverly Hills, CA: Sage.

Browning, C. 1974. Population and urbanized area growth of megalopolis, 1950–1970. In *Studies in geography,* No. 7. Chapel Hill: University of North Carolina.

Castells, M. 1989. *The informational city.* Oxford, U.K.: Blackwell.

Dorgan, M., and J. Kasarda. 1988. *The metropolis era.* Newbury Park, CA: Sage.

Dunn, E. 1983. *The development of the US urban system.* Baltimore: Johns Hopkins University Press.

Florida, R. 2002. *The rise of the creative class,* 2nd rev. ed. New York: Basic Books.

Frey, W. H., and A. Speare. 1988. *Regional and metropolitan growth and decline in the United States.* New York: Russell Sage.

Garreau, J. 1991. *Edge city: Life on the new frontier.* New York: Doubleday.

Gottmann, J. 1961. *Megalopolis: The urbanized northeastern seaboard of the United States.* New York: Twentieth Century Fund.

———. 1976. Megalopolitan systems around the world. *Ekistics* 243:109–13.

———. 1987. *Megalopolis revisited: 25 years later.* Baltimore: University of Maryland Institute for Urban Studies.

Harvey, D. 1989. *The urban experience.* Baltimore: Johns Hopkins University Press.

Johnston, R. 1982. *The American urban system.* New York: St. Martins Press.

Kahn, H., and A. Wiener. 1967. *The year 2000.* New York: Macmillan.

Knox, P., ed. 1993. *The restless urban landscape.* New York: Prentice Hall.

Mayer, H. 1969. *The spatial expression of urban growth,* Commission on College Geography Resource Paper 7. Washington, DC: Association of American Geographers.

Morrill, R., J. Cromartie, and G. Hart. 1999. Metropolitan, urban and rural commuting areas: Toward a better depiction of the United States settlement system. *Urban Studies* 20:727–48.

Muller, P. 1981. *Contemporary suburban America.* Englewood Cliffs, NJ: Prentice Hall.

Noyelle, T., and T. Stanback. 1984. *The economic transformation of American cities.* Totowa, NJ: Rowman and Allanhel.

Orfield, M. 2002. *American metropolitics: The new suburban realities.* Washington, DC: Brookings.

Peirce, N. 1993. *Citistates. How urban America can prosper in a competitive world.* Washington, DC: Seven Locks Press.

Regional Plan Association. 1967. *The region's growth.* New York: Regional Plan Association.

Sassen, S. 1991. *The global city: New York, London, Tokyo.* Princeton, NJ: Princeton University Press.

Scott, A., ed. 2001. *Global city regions.* Oxford, U.K.: Oxford University Press.

Smith, N. 1986. *Gentrification of the city.* Boston: Allen and Unwin.

Soja, E. 2000. *Postmetropolis: A critical study of cities and regions.* Oxford, U.K.: Blackwell.

Stanback, T. 1991. *The new suburbanization: Challenge to the central city.* Boulder, CO: Westview Press.

Wheeler, J., Y. Aoyama, and B. Warf, eds. 2000. *Cities in the telecommunications age: The fracturing of geographies.* New York: Routledge.

RICHARD MORRILL is Professor Emeritus in the Department of Geography of the University of Washington, Seattle WA 98195. E-mail. morrill@u.washington.edu. His research interests cover most of human geography.

UNIT 2

Human–Environment Relations

Unit Selections

Key Points to Consider

- What are the long-range implications of atmospheric pollution? Explain the greenhouse effect.

- How can the problem of regional transfer of pollutants be solved?

- The manufacture of goods needed by humans produces pollutants that degrade the environment. How can this dilemma be solved?

- Where in the world are there serious problems of desertification and drought? Why are these areas increasing in size?

- How are you as an individual related to the land? Does urban sprawl concern you? Explain.

- How will energy demands be satisfied in the future as more countries industrialize?

- Why should imbalances in urban and rural viability be of concern to you?

Student Website
www.mhcls.com

Internet References

Alliance for Global Sustainability (AGS)
http://www.global-sustainability.org

California Climate Change Portal
http://www.climatechange.ca.gov

Human Geography
http://www.le.ac.uk/gg/cti/

The North-South Institute
http://www.nsi-ins.ca

United Nations Environment Programme (UNEP)
http://www.unep.ch

US Global Change Research Program
http://www.usgcrp.gov

World Health Organization
http://www.who.int

The home of humankind is Earth's surface and the thin layer of atmosphere enveloping it. Here the human populace has struggled over time to change the physical setting and to create the telltale signs of occupation. Humankind has greatly modified Earth's surface to suit its purposes. At the same time, we have been greatly influenced by the very environment that we have worked to change. This basic relationship of humans and land is important in geography.

Geographers observe, study, and analyze the ways in which human occupants of Earth have interacted with the physical environment. This unit presents a number of articles that illustrate the theme of human–environment relationships. In some cases, the association of humans and the physical world has been mutually beneficial; in others, environmental degradation has been the result.

At the present time, the potential for major modifications of Earth's surface and atmosphere is greater than at any other time in history. It is crucial that the environmental consequences of these modifications be clearly understood before such efforts are undertaken.

"A Great Wall of Waste" draws attention to the enormity of environmental pollution as China industrializes. The positive combination of increased rainfall and sound agricultural practices are reclaiming significant tree cover in desertified Niger. The next article deals with the balance between industrial development and maintaining the Amazon rainforest.

"Why It's Time for a 'Green New Deal'" shines yet another light on the need for environmental awareness. Human impact on the oceans is the subject of the next article. The fate of the polar bear is considered in "Polar Distress." Finally, the case is made that China can learn a great deal about avoiding environmental degradation from Japan.

This unit provides a small sample of the many ways in which humans interact with the environment. The outcomes of these interactions may be positive or negative. They may enhance the position of humankind and protect the environment, or they may do just the opposite. We human beings are the guardians of the physical world. We have it in our power to protect, to neglect, or to destroy.

© Nigel Hicks/Alamy

A Great Wall of Waste

China is slowly starting to tackle its huge pollution problems.

Plugging a cigarette into his mouth, He Shouming runs a nicotine-stained fingernail down a list of registered deaths in Shangba, dubbed "cancer village" by the locals. The Communist Party official in this cluster of tiny hamlets of 3,300 people in northern Guangdong province, he concludes that almost half the 11 deaths among his neighbours this year, and 14 of the 31 last year, were due to cancer.

Mr He blames Dabaoshan, a nearby mineral mine owned by the Guangdong provincial government, and a host of smaller private mines for spewing toxic waste into the local rivers, raising lead levels to 44 times permitted rates. Walking around the village, the water in the streams is indeed an alarming rust-red. A rice farmer complains of itchy legs from the paddies, and his wife needs a new kettle each month because the water corrodes metal. "Put a duck in this water and it would die in two days," declares Mr He.

Poisons from the mines are also killing the village's economy, which depends on clean water to irrigate its crops, says Mr He. Rice yields are one-third of the national average and nobody wants to buy the crop. Annual incomes here have been stuck at less than 1,500 yuan ($180) per person for a decade, almost three times lower than the average in Guangdong province. The solution to Shangba's nightmare would be a local reservoir, but that idea was abandoned after various tiers of government squabbled over the 8.4m yuan cost.

Some 200 km (124 miles) farther south and several decades into the future sits the Taihe landfill plant. Built for 540m yuan by Onyx, a waste-management company that is part of Veolia, a French utility, it has handled all of Guangzhou city's solid waste for the past two years. Each hour 140 trucks snake into the site, bringing 7,000 tonnes of rubbish a day from the 9.9m inhabitants of Guangdong's capital. In October delegates from 300 other municipalities will visit Taihe, promoted by central government as a role model of technology.

Smart cards record each truck's load, since Onyx charges by weight. Unrippable German fabric lines the crater into which the waste is dumped, stopping leachate—a toxic black liquid—from leaking into the groundwater, as it does at almost all Chinese-run sites. Most landfill in China is wet (solid rubbish, such as old TVs, is scavenged), and the Taihe plant collects a full 1,300 tonnes of the black liquid daily. Chemical and filtration systems to neutralise it are its biggest cost. Expensive too is the extraction equipment to gather another by-product, methane

gas, which Onyx plans to feed into generators that will supply electricity to the local grid. Finally, the waste is topped off with plastic caps, deodorised and landscaped, while a crystal-clear fountain at the entrance tinkles with the cleaned-up leachate.

The extremes represented by Shangba and Taihe explain why it is difficult to get an accurate picture of China's pollution. In a country where data are untrustworthy, corruption rife and the business climate for foreigners unpredictable, neither the cause of Shangba's problems nor the smooth efficiency of Taihe are necessarily what they seem. As with many other aspects of China's economic development, rapid progress and bold experiments in some areas are balanced by bureaucratic rigidity and stagnation in others.

Certainly, awareness of China's environmental problems is rising among policymakers at the highest level—reflected in a new package of right-sounding initiatives like a "green GDP" indicator to account for environmental costs. So is the pressure, both internal and international, to fix them. But while all developing economies face this issue, there are historical, political and institutional reasons why it will be a long and complicated process in China. There is some cause for optimism, not least an influx of foreign technology and capital. But progress on pollution is unlikely to be as rapid or uniform as the government and environmentalists desire.

Nor should it necessarily be. China's need to lift so many people out of poverty (the country's average annual income per head has only just breached $1,000), holds the edge over long-term considerations like sustainable development. The priorities

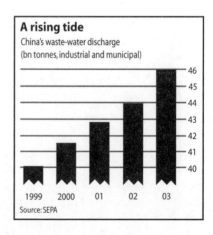

A rising tide
China's waste-water discharge
(bn tonnes, industrial and municipal)

1999   2000   01   02   03
Source: SEPA

of environmental activists, both foreign and Chinese, almost never reflect this. Greenpeace lobbies for China to invest in wind farms, an unrealistic answer to the country's power needs, while environmentalists from rich countries naively tell aspiring Chinese to eschew their new cars and air-conditioners.

Nor Any Drop to Drink

That is not to deny the huge scale of China's environmental challenges. Water and waste pollution is the single most serious issue. Pan Yue, deputy head of the State Environmental Protection Administration (SEPA), the country's environmental watchdog ministry, calls it "the bottleneck constraining economic growth in China." Per head, China's water resources are among the lowest in the world and concentrated in the south, so that the north and west experience regular droughts. Inadequate investments in supply and treatment infrastructure means that even where water is not scarce, it is rarely clean. Around half the population, or 600 m people, have water supplies that are contaminated by animal and human waste.

In late July an environmental disaster occurred on the Huai river, one of China's seven big rivers. A 133 km-long black and brown plume swept along the river killing millions of fish and devastating wildlife. According to Mr Pan, the catastrophe occurred because too much water had been taken from the river system, reducing its ability to clean itself. Others say that numerous factories dump untreated waste directly into the water.

As for used water, with a national daily sewage rate of around 3.7 billion tonnes, China would need 10,000 waste-water treatment plants costing some $48 billion just to achieve a 50% treatment rate, according to Frost & Sullivan, a consultancy. SEPA found over 70% of the water in five of China's seven major river systems was unsuitable for human contact. As more people move into cities, the problem of household waste is becoming severe. Only 20% of China's 168 m tonnes of solid waste per year is properly disposed of.

The air is not much better. "If I work in your Beijing, I would shorten my life at least five years," Zhu Rongji told city officials when he was prime minister in 1999. According to the World Bank, China has 16 of the world's 20 most polluted cities. Estimates suggest that 300,000 people a year die prematurely from respiratory diseases.

The main reason is that around 70% of China's mushrooming energy needs are supplied by coal-fired power stations, compared with 50% in America. Combined with the still widespread use of coal burners to heat homes, China has the world's highest emissions of sulphur dioxide and a quarter of the country endures acid rain. In 2002, SEPA found that the air quality in almost two-thirds of 300 cities it tested failed World Health Organisation standards—yet emissions from rocketing car ownership are only just becoming an issue. Hopes that China will "leapfrog" the West with super-green cars are naive, since dirty fuel messes up clean engines and the high cost of new cars keeps old ones on the road. Sun Jian, the second-ranking official at Shanghai's environmental protection bureau, estimates that 70% of Shanghai's 1 m cars do not even reach the oldest European emission standards.

Farmland erosion and desertification resulted in Beijing being hit with 11 sandstorms in 2000, prompting Mr Zhu to wonder whether the advancing desert might force him to relocate the capital. A year later, the yellow dust clouds were so extensive that they raised complaints in South Korea and Japan and travelled as far as America. A partial logging ban and massive replanting appear to have reversed China's deforestation, but its grass and agricultural land continue to shrink.

Adding it all up, the World Bank concludes that pollution is costing China an annual 8–12% of its $1.4 trillion GDP in direct damage, such as the impact on crops of acid rain, medical bills, lost work from illness, money spent on disaster relief following floods and the implied costs of resource depletion. With health costs escalating, that figure will increase, giving rise to some grim prognoses that growth itself will be undermined. "Ignored for decades, even centuries, China's environmental problems have the potential to bring the country to its knees economically," argues Elizabeth Economy, author of "The River Runs Black", a new book on China's pollution.

SEPA's Mr Pan is gloomier still: "Our natural resources will soon be unable to support our population." His predecessor Qu Geping, the first head of China's National Environmental Protection Agency (SEPA's forerunner) in 1985, believes that while the official goal of quadrupling 2002 GDP by 2020 can be "healthily achieved", if nothing is done about the environment, economic growth could grind to a halt.

But China's relationship with its environment has long been uneasy. For centuries, the country's rulers subjugated their surroundings rather than attempting to live in harmony with them. Mao declared that man must "conquer nature and thus attain freedom from nature". In the past two decades, the toll extracted by China's manufacturing-led development and the sheer scale of its 9%-a-year economic expansion has only increased.

From Conquest to Nurture

This has spurred the government into belated action. In 1998, Mr Zhu elevated SEPA to ministerial rank and three years later the 10th Five-year Plan for Environmental Protection set ambitious emission-reduction targets and boosted environmental spending to 700 billion yuan ($85 billion) for 2001–05—equivalent to 1.3% of GDP, up from 0.8% in the early 1990s (though still

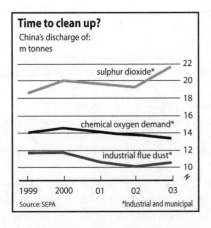

Time to clean up?
China's discharge of:
m tonnes

sulphur dioxide*

chemical oxygen demand*

industrial flue dust*

1999 2000 01 02 03

Source: SEPA *Industrial and municipal

below the 2% suggested by the World Bank). A legal framework has been created. And the rhetoric has changed too, with Hu Jintao and Wen Jiabao, the current president and prime minister, now stressing balanced development rather than all-out economic growth.

Beijing's good intentions, however, have so far had only limited impact, thanks to the vast, decentralised bureaucracy through which it is forced to govern such a huge country. As Ken Lieberthal, a China expert at the University of Michigan, explains: "Much of the environmental energy generated at the national level dissipates as it diffuses through the multi-layered state structure, producing outcomes that have little concrete effect."

SEPA, the government's chosen weapon in the fight against pollution, is under-resourced despite its enhanced status, with little money and just 300 central staff. In the capital, it must battle for influence with other agencies, such as the Construction Ministry that handles water and sewage treatment. Bureaucratic rivalries mean there is no co-operation and no sharing of the (often patchy) data that are collected with limited funds, observes Bruce Murray, the Asian Development Bank's representative in China.

Around the country, SEPA's branches, known as Environmental Protection Bureaus, are supposed to monitor pollution, enforce standards and collect fines. But they are more in thrall to local governments—whose priorities are to maintain growth and employment in their jurisdiction—than to head office in Beijing. It is no rarity, therefore, to find a bureau imposing a fine on a dirty local enterprise (thus fulfilling its duty), but then passing the money on to the local administration, which refunds it to the company via a tax break. "The environmental management system needs real reform," says Ma Jun, an environmental scholar. "The bureaus depend on the local government for their salaries and pensions. How can they enforce regulations against the local government?" Mr Pan complains that SEPA cannot effectively push through central edicts because it does not directly employ environmental personnel at the local level. Mr Sun at the Shanghai bureau says that SEPA has given him only 300 people with which to police 20,000 factories.

Make Polluters Pay

SEPA's impotence is one reason why penalties, even when it can impose them, remain laughably light. Mr Sun says the maximum he can fine a polluting company in Shanghai—a model city when it comes to the environment—is 100,000 yuan or about $12,000. But just as fundamental is that China lacks an understanding of the concept that the polluter should pay. "The legacy of the old, centrally planned economy is that electricity and water are treated as free goods or goods to be provided at minimal cost," says the ADB's Mr Murray. Since the utilities cannot pass on the costs of cleaner water or lower power-station emissions to consumers, they fight any drive for higher standards and conservation tooth and nail. Even the central government is unwilling to impose price rises in basic services that could spark public unrest.

Water is an example. While customer tariffs have been raised in showcase cities, such as Beijing and Dalian in the northeast, water remains stunningly cheap in China. According to the World Bank, water for agriculture, which makes up three-quarters of the total used, is priced at 0.03 yuan (0.4 cents) per cubic metre, or about 40% of cost. More than half is lost in leaky irrigation systems. Meanwhile, the cost of more modern services, such as Guangzhou's solid-waste disposal, is entirely borne by the government.

Without the introduction of realistic pricing, China will not be able to afford to clean up its pollution, particularly the cost of enough foreign technology. Yet a system allocating the costs to the polluter will be hard to introduce and enforce. Even in Hong Kong, the territory's environment minister Sarah Liao concedes there is no tradition of having consumers bear the full costs of environmental regulations.

Ms Liao can also testify to the mainland's ambivalent attitude when it comes to letting outsiders help. She started looking into how the Pearl River delta's pollution was affecting Hong Kong back in 1999, but her requests to start monitoring emissions were repeatedly rebuffed even when she offered to pay for the equipment. Data collection finally started this year. For Thames Water, a British utility that is now a part of Germany's RWE, the experience was much worse. In June, Thames pulled out of a $73 m advanced waste-water treatment plant it had built and was running in Shanghai, after the central government ruled that the fixed annual 15% return it had negotiated was now illegal.

There is no need to be unremittingly gloomy about China's environment, nevertheless. As developing countries get richer, they tend to pollute less. Nationally in China, discharges of chemical oxygen have declined over the past three years, those of industrial dust have stabilised and sulphur-dioxide emissions had been on the downtrend until 2003 when energy shortages increased demand for sulphurous coal (see charts). Most east-coast cities are enjoying more sunny days and the pollution load in the rivers is falling. Environmentally, in many places, China may have passed its nadir.

The government is increasing environmental spending and the more concerned attitude of the top leadership could filter down the hierarchy if the performance of officials starts being measured partly on environmental criteria, as Mr Qu hints it might. But the bigger incentive is that Beijing is under pressure to do more, partly from domestic public opinion. As urban Chinese see their material wealth increase, more are caring about the environment, while the concerns of the poor are increasingly being channelled by green non-governmental organisations. Though these remain extremely weak—few have more than a handful of members and all need government affiliation— Mr Wen said recently he suspended plans for the construction of 13 dams along the Nu river in Yunnan province partly because of the concerns outlined by such groups.

External pressure is even greater. Despite reservations, foreign companies are flocking to China, scenting a fast-growing market for their environmental technologies and skills. International agencies are tying funds to environmental criteria, while foreign governments are beginning to complain about China's dust storms and greenhouse-gas emissions. All this will help spread best practices. Beijing is fast cleaning up ahead of the 2008 Olympics, moving out factories and introducing clean-vehicle technology: a new premium is being placed on global respectability.

Of course, environmental problems and their huge costs will dog China for many years. In a country where the public is not free to speak, too many courts are toothless and environmental groups remain on a tight leash, it will be hard to know if the government's avowedly green policies are being implemented. But China deserves credit for its attempts to clean itself up. The balance between sustainable development and economic growth will have to be continuously adjusted in the future. Right now, China is probably moving in the right direction.

In Niger, Trees and Crops Turn Back the Desert

A Poor African Nation Uses a Simple Mix to Grow Greener

LYDIA POLGREEN

In this dust-choked region, long seen as an increasingly barren wasteland decaying into desert, millions of trees are flourishing, thanks in part to poor farmers whose simple methods cost little or nothing at all.

Better conservation and improved rainfall have led to at least 7.4 million newly tree-covered acres in Niger, researchers have found, achieved largely without relying on the large-scale planting of trees or other expensive methods often advocated by African politicians and aid groups for halting desertification, the process by which soil loses its fertility.

Recent studies of vegetation patterns, based on detailed satellite images and on-the-ground inventories of trees, have found that Niger, a place of persistent hunger and deprivation, has recently added millions of new trees and is now far greener than it was 30 years ago.

These gains, moreover, have come at a time when the population of Niger has exploded, confounding the conventional wisdom that population growth leads to the loss of trees and accelerates land degradation, scientists studying Niger say. The vegetation is densest, researchers have found, in some of the most densely populated regions of the country.

"The general picture of the Sahel is much less bleak than we tend to assume," said Chris P. Reij, a soil conservationist who has been working in the region for more than 30 years and helped lead a study published last summer on Niger's vegetation patterns. "Niger was for us an enormous surprise."

About 20 years ago, farmers like Ibrahim Danjimo realized something terrible was happening to their fields.

"We look around, all the trees were far from the village," said Mr. Danjimo, a farmer in his 40s who has been working the rocky, sandy soil of this tiny village since he was a child. "Suddenly, the trees were all gone."

Fierce winds were carrying off the topsoil of their once-productive land. Sand dunes threatened to swallow huts. Wells ran dry. Across the Sahel, a semiarid belt that spans Africa just below the Sahara and is home to some of the poorest people on earth, a cataclysm was unfolding.

Severe drought in the 1970s and 80s, coupled with a population explosion and destructive farming and livestock practices, was denuding vast swaths of land. The desert seemed determined to swallow everything.

So Mr. Danjimo and other farmers in Guidan Bakoye took a small but radical step. No longer would they clear the saplings from their fields before planting, as they had for generations. Instead they would protect and nurture them, carefully plowing around them when sowing millet, sorghum, peanuts and beans.

Today, the success in growing new trees suggests that the harm to much of the Sahel may not have been permanent, but a temporary loss of fertility. The evidence, scientists say, demonstrates how relatively small changes in human behavior can transform the regional ecology, restoring its biodiversity and productivity.

In Niger's case, farmers began protecting trees just as rainfall levels began to rise again after the droughts in the 1970s and 80s. Another change was the way trees were regarded by law. From colonial times, all trees in Niger had been regarded as the property of the state, which gave farmers little incentive to protect them. Trees were chopped for firewood or construction without regard to the environmental costs. Government foresters were supposed to make sure the trees were properly managed, but there were not enough of them to police a country nearly twice the size of Texas.

But over time, farmers began to regard the trees in their fields as their property, and in recent years the government has recognized the benefits of that outlook by allowing individuals to own trees. Farmers make money from the trees by selling branches, pods, fruit and bark. Because those sales are more lucrative over time than simply chopping down the tree for firewood, the farmers preserve them.

The greening began in the mid-1980s, Dr. Reij said, "and every time we went back to Niger, the scale increased."

"The density is so spectacular," he said.

Mahamane Larwanou, a forestry expert at the University of Niamey in Niger's capital, said the regrowth of trees had transformed rural life in Niger.

"The benefits are so many it is really astonishing," Dr. Larwanou said. "The farmers can sell the branches for money. They can feed the pods as fodder to their animals. They can sell or eat the leaves. They can sell and eat the fruits. Trees are so valuable to farmers, so they protect them."

They also have extraordinary ecological benefits. Their roots fix the soil in place, preventing it from being carried off with the fierce Sahelian winds and preserving arable land. The roots also help hold water in the ground, rather than letting it run off across rocky, barren fields into gullies where it floods villages and destroys crops.

One tree in particular, the Faidherbia albida, known locally as the gao tree, is particularly essential. It is a nitrogen-fixing tree, which helps fertilize the soil. Its leaves fall off during the rainy season, which means it does not compete with crops for water, sun or nutrients during the growing period. The leaves themselves become organic fertilizer when they fall.

"This tree is perfectly adapted for farming in the Sahel," said Dr. Larwanou. "Yet it had all but disappeared from the region."

That is because for generations local farmers had simply cleared their fields of all vegetation, including trees, before sowing neat rows of sorghum, millet, peanuts and beans. When a field became less productive, the farmer would move on to another.

Wresting subsistence for 13 million people from Niger's fragile ecology is something akin to a puzzle. Less than 12 percent of its land can be cultivated, and much of that is densely populated. Yet 90 percent of Niger's people live off agriculture, cultivating a semi-arid strip along the southern edge of the country.

Farmers here practice mostly rain-fed agriculture with few tools and no machinery, making survival precarious even in so-called normal times. But when the rains and harvest fall short, hunger returns with a particular vengeance, as it did in 2005 during the nation's worst food crisis in a generation.

Making matters worse, Niger's population has doubled in the last 20 years. Each woman bears about seven children, giving the country one of the highest growth rates in the world.

The regrowth of trees increases the income of rural farmers, cushioning the boom and bust cycle of farming and herding.

Ibrahim Idy, a farmer in Dahirou, a village in the Zinder region, has 20 baobab trees in his fields. Selling the leaves and fruit brings him about $300 a year in additional income. He has used that money to buy a motorized pump to draw water from his well to irrigate his cabbage and lettuce fields. His neighbors, who have fewer baobabs, use their children to draw water and dig and direct the mud channels that send water coursing to the beds. While their children work the fields, Mr. Idy's children attend school.

In some regions, swaths of land that had fallen out of use are being reclaimed, using labor-intensive but inexpensive techniques.

In the village of Koloma Baba, in the Tahoua region just south of the desert's edge, a group of widows have reclaimed fields once thought forever barren. The women dig small pits in plots of land as hard as asphalt. They place a shovelful of manure in the pits, then wait for rain. The pits help the water and manure stay in the soil and regenerate its fertility, said Dr. Larwanou. Over time, with careful tending, the land can regain its ability to produce crops. In this manner, more than 600,000 acres of land have been reclaimed, according to researchers.

Still, Koloma Baba also demonstrates the limits of this fragile ecosystem, where disaster is always one missed rainfall away. Most able-bodied young men migrate to Nigeria and beyond in search of work, supporting their families with remittances. The women struggle to eke a modest crop from their fields.

"I produce enough to eat, but nothing more," said Hadijatou Moussa, a widow in Koloma Baba.

The women have managed to grow trees on their fields as well, but have not seen much profit from them. People come and chop their branches without permission, and a village committee that is supposed to enforce the rights of farmers to their trees does not take action against poachers.

Such problems raise the question of whether the success of some of Niger's farmers can be replicated on a larger scale, across the Sahel. While Niger's experience of greening on a vast scale is unique, scientists say, smaller tracts of land have been revived in other countries.

"It really requires the effort of the whole community," said Dr. Larwanou. "If farmers don't take action themselves and the community doesn't support it, farmer-managed regeneration cannot work."

In a poor country, farmers create a low-cost, low-tech success.

Moussa Bara, the chief of Dansaga, a village in the Ague region of Niger, where the regeneration has been a huge success, said the village has benefited enormously from the regrowth of trees. He said not a single child died of malnutrition in the hunger crisis that gripped Niger in 2005, largely because of extra income from selling firewood. Still, he said, the village has too many mouths to feed.

"We are many and the land is small," he explained, bouncing on his lap a little boy named Ibrahim, the youngest of his 17 children by his three wives.

Climate change is another looming threat. Kerry H. Cook, a professor of atmospheric science at Cornell University, said that improved rains in the Sahel are most likely a result of natural climate variability from decade to decade, and that while the trend is positive, the rains have not entirely recovered to what they were in the 1950s.

The Sahel, like other parts of Africa, has experienced big swings in rainfall in recent years. Severe droughts in eastern and southern Africa have led to serious hunger crises in the past five years, and a drop in precipitation in Niger in 2005 contributed to the food crisis here that year.

Dr. Cook's long-term projections, based on a variety of climate models, point to longer and more frequent dry periods in the Sahel, caused by rising temperatures in the Gulf of Guinea.

"This is the place in the world that just stands out for having vulnerability for drought," she said.

Still, more trees mean that Niger's people are in a better position to withstand whatever changes the climate might bring. "This is something the farmers control, and something they do for themselves," said Dr. Larwanou. "It demonstrates that with a little effort and foresight, you can reduce poverty in the Sahel. It is not impossible or hopeless, and does not have to cost a lot of money. It can be done."

Whither the World's Last Forest?

Brazil bets that it can save the Amazon wilderness while tapping its riches.

MARK LONDON AND BRIAN KELLY

The flame on the horizon is startling, a tight orange cone shimmering over the tree line. After flying for almost two hours southwest from Manaus with nothing but trees and an occasional snaking brown river underneath, any sign of civilization is satisfying. The fire's source becomes clear as we approach: a sprawling series of white chimneys, part of a high-tech industrial complex that looks like a secret military installation. An army of workers in orange jumpsuits moves through a maze of pipes and steel towers and low squat buildings. We hadn't seen a town for hundreds of miles in any direction, not even a road, except for the spine of black pavement we spotted as we approached this clearing. A wildcatter from Oklahoma exploring for oil in the Peruvian Amazon once said to us, "As a general rule, you have to remember the good Lord was a fine man, but he picked some godawful places to put oil." This was one of them.

The oil and gas field at the headwaters of the Urucu River lies almost dead center in the South American continent, surrounded by primary rain forest for hundreds of miles in all directions. If there were a part of the Amazon that even the most worrisome environmentalist considered impenetrable, this would be it.

It's estimated that there are at least 100 billion cubic meters of gas and 18 million barrels of oil in the Urucu region. "This is not Saudi Arabia, but for Brazil it will be very helpful," said Ronaldo Coelho, who manages the site for Petrobras, the state-owned oil company.

The hydrocarbons are high quality and easily recoverable. The crude is unusually pure, bubbling out of the wellhead like espresso. "You could practically strain this through your handkerchief and put it in your gas tank," said Coelho as he rubbed some between his fingers. "The only issue is how to get it out of this site to a market. And that's a political problem, not a technical one." A big political problem.

Whenever an access route has been created in the Amazon, a spontaneous influx of immigrants hungry for land has emerged. Environmentalists see the gas and oil finds as a death blow to the remote western jungle, fearing that pipelines to Manaus and Porto Velho in the southwestern Amazon will open a seam of entry to empty forest and protected Indian lands, clearing the way for a torrent of loggers, miners, and cattle ranchers.

The controversy over the pipelines—along with other burgeoning industries such as cattle ranching, soy farming, and iron mining—has profoundly changed the traditional debate about how to manage the Amazon—or, as many environmentalists would see it, how to save the Amazon. The construction of these pipelines will alter the rain forest but will also generate energy for millions of people. Nearly 2 million people live in Manaus alone, and they need energy. Blackouts rotate through the city daily. Lack of energy has retarded factory construction, holding back employment expansion. When Brazilian President Lula da Silva approved the pipeline to Manaus in the spring of 2004, he said, "If people want development that preserves the environment, we have to have energy. It's no good people saying the Amazon has to be the sanctuary of humanity and forget there are 20 million people living there."

The Amazon is not, and never has been, a pristine wilderness that could be fenced and preserved as an intact ecosystem. Increasingly, it is proving to be a resource-rich region of a continent that desperately needs to grow. Brazil, which contains most of the Amazon basin, is under particular pressure as it tries to reconcile its great disparities between rich and poor. And there's a voracious market for the goods, whether it's the Chinese buying steel or the Europeans buying soybeans. At the same time, the vast basin of freshwater and forest is a global feature of such magnitude that its destruction will only help tip a fragile global climate further over the edge. The hard question facing the various governments and organizations with a stake in the outcome is whether some development can prevent a lot of deforestation.

The hard question facing Brazil is whether some development can prevent a lot of deforestation.

Every year a chunk of forest equivalent to an average-size U.S. state disappears from the Amazon. In the year ending August 2004, 16,236 square miles, about twice the size of

Massachusetts, were deforested. According to Conservation International, that represents between 1.1 billion and 1.4 billion trees of 4 inches or more in diameter. This deforestation took place during a time of heightened environmentalism in Brazil, during a robust return to democracy when a comprehensive body of laws protecting the Amazon had been enacted and supported by broad enforcement powers—though often, not the enforcement itself. The reaction of the Brazilian government and nongovernmental organizations to these annual figures can be summarized by the Yogi Berra quote, "It's like déjà vu all over again." The so-called experts annually express "shock and surprise" at the figures. The shock subsides, then reappears the following spring. Fingers point at the culprit du jour—the cattle ranchers in some years, or the soy farmers, or the migration of small families clearing homesteads. Loggers, miners, and ranchers get denounced regularly. And in response, the government usually sets aside another national park equivalent in size to a small American state. A federal department's budget gets increased by more than $100 million, at least publicly. A government official sometimes resigns. Nongovernmental organizations use the statistics in their annual pleas for contributions. The *New York Times* writes an editorial reminding Brazil that "the rain forest is not a commodity to be exploited for private gain." The *Economist* chides Brazil for its institutions, which are "weak, poorly coordinated, and prone to corruption and influence-peddling." But from one year to another, the process repeats itself and the Amazon shrinks. When we first traveled here in 1980, about 3 percent had been deforested. Today, more than 20 percent is gone.

That number needs some interpretation. Compared with the dire predictions of 25 years ago—that most of the forest would be destroyed by now—it actually looks good. And there's widespread acceptance that even more forest inevitably will be cleared. The problem is how that clearing is managed. Now it is haphazard and uncontrollable. The emerging consensus, at least among the key decision makers in Brazil, is that the solution is more development, not less. The argument is that development means civilization, which brings the resources to create better economic incentives and to enforce the laws. The downside is that if the Brazilian strategy doesn't work, it will be too late to change course.

"You have to understand that deforestation is not just about the environment," says Everton Vargas, the top environmental strategist for the Brazilian Foreign Ministry. "Deforestation is an economic issue. It will not be avoided simply by saying, 'Don't cut the trees.' You have to say, 'Here's why you don't have to cut the trees.'"

Finding those incentives and making them work is a job that keeps Eduardo Braga up at night. The governor of the state of Amazonas is one of the most important decision makers when it comes to the future of the Amazon. After a long day at his office in Manaus, he slumps from the stress of trying to administer a territory as vast as the land between Chicago and Juneau, Alaska. His outer office is filled with small-town mayors who have traveled days just to meet with him. "I am constantly tired," he confides. "There is so much to do. So much space to cover."

His optimism comes from two serendipities that he inherited on taking office six years ago. The first can be found in the Zona Franca, an incongruous sprawl of modern manufacturing plants that rings the outskirts of Manaus, which was the capital of the turn-of-the-century rubber boom and now has turned into a mix of glassy high-rise condos, suburban housing tracts, and fetid Latin American slums set amid majestic but peeling colonial buildings. The tax-free Zona Franca takes in parts from international brands like Honda and Nokia and ships out finished motorcycles and cellphones. The other windfall is the natural gas discovery.

"Gas changes everything for us. It will give us the energy to allow industry to grow in Manaus. It will give us the energy in the small towns to improve their quality of life. Gas will give us the money to do other things, to improve social services here and to have programs to develop the rest of the state in a way that protects the environment."

He plans to create a network of family farms and supporting towns to provide a bulwark against uncontrolled development. "It's inevitable that people are going to invade these areas," he says. Braga sees two choices for Amazonas on its southern flank: spillover development and the resulting anarchy and violence endemic elsewhere in the Amazon, or some semblance of civil society. "If we have roads, we can put IBAMA [the environmental protection agency] there. We can put government agencies there. We can put schools there. We can put health centers there. We can create conditions for family farms that are clearly demarcated and where people can make a living. You think that no controls means no people? No controls means that people just invade the land and do what they want. The people already are there, and we can't leave them behind like a bag of trash. We need to connect them."

But Braga also knows that the cycle of development, once started, cannot be stopped. It is based on an economic, not ecological, choice. "I understand that the small farms eventually will sell out to the big farms, and then you end up with major agricultural interests and small people in search of land. As long as using the land brings more material benefits to people than not using the land, we don't really have much chance. I hope to break the cycle."

The governor of Amazonas state is promoting a Green Free-Trade Zone and the creation of "certified" forests.

Braga calls the program the Zona Franca Verde, or the Green Free-Trade Zone. He's promoting a range of local products to help create stable communities: guarana berries, which make a popular soft drink; jute fibers; fish farms. This comes under the rubric of "sustainable development," an ill-defined buzzword of the international development community that has so far shown mixed to disappointing results elsewhere in the Amazon.

Much more important may be large-scale forest management through the creation of so-called certified forests. Braga wants to lease timber concessions to big companies that would practice sustainable forestry by carefully harvesting and replanting trees. The companies in turn sell their lumber to U.S. and European importers who agree to buy only certified wood.

And his ultimate goal is to tie the Amazon into some sort of international carbon market that, by putting a price on the carbon contained in the trees, would create an incentive to not cut and burn them. Carbon-trading markets exist in Europe and the United States on a mostly experimental basis. If they became global, the rights to billions of trees—whether in the Amazon or other endangered forests in South Asia and Africa—could become quite valuable.

That may be far off, but Braga sees a much more immediate possibility of bringing foreign investment to Amazonas as a way to break out of the traditional Third World cycle of exporting low-priced raw materials to advanced factories in the developed world. "People want to save the forest? They want to help?"

Braga asks. "We need resources to establish these programs. Maybe Home Depot wants to build a factory here and will buy only certified wood. Let us add value here. Then we can take those profits and return them to the people." The area of Carajás has the world's greatest iron ore reserves, but there are no steel mills on-site. Trombetas has one of the world's greatest bauxite mines, but there are no aluminum mills in proximity. "It's frustrating," Braga says. "It's frustrating when the Kyoto Protocol does nothing to help us. It's frustrating when we try to open markets to products and we can't get the investment we need to support the production."

So for now, the Brazilians have decided to try to forgo the pleadings and promises of the international community and take their chances on promoting aggressive development while hoping they can control its effects. "Other countries just are going to have to trust us to take care of the Amazon," says Everton Vargas. "That's the way it'll have to be."

Why It's Time for a "Green New Deal"

**As the world faces economic turmoil, cleaner energy
can create jobs and reignite global growth.**

CHRISTOPHER DICKEY AND TRACY MCNICOLL

In rented offices on a quiet side street in Paris, not far from the Eiffel Tower, analysts for the International Energy Agency spend long days and nights crunching numbers about oil production and greenhouse-gas emissions. They're the staid, sober global accountants who watch over the power supply for the 30 rich countries that are members of the Organization for Economic Cooperation and Development, and their many reports are dry and technical. But lately, the group's pronouncements have taken on more ominous overtones. With a sense of urgency bordering on desperation, the IEA has begun calling for radical changes in the way the world drives its cars, its factories and, indeed, the global economy. This month the agency will issue a collection of comprehensive reports declaring that "a global revolution is needed in ways that energy is supplied and used."

That kind of rhetoric has become familiar to U.S. voters, who've spent months listening to both presidential candidates tout their energy plans. Barack Obama has promised to "strategically invest" $150 billion over 10 years to build a clean-energy economy, one that will create 5 million new green jobs. While John McCain has offered slightly fewer specifics, he's promoted an "all of the above" strategy that focuses more on nuclear energy and drilling for more oil. "The U.S. must become a leader in a new international green economy," McCain has said.

Starting this week, one of those candidates will have a chance to make good on those promises. On the surface, that opportunity could hardly come at a worse time. Around the globe, the financial crisis plunging the world into recession has caused a wave of doubts, second-guessing and backsliding among many political and business leaders who've pledged allegiance to a green agenda. Italy and several Eastern European states have threatened to sink previously agreed-upon European Union initiatives. U.S. venture capitalists are getting cautious about investing in clean-tech projects. China's leaders, after what seemed a crisis of environmental conscience during the Beijing Olympics, may reverse course as they see their economic growth drop into single digits.

But there are also powerful voices being raised amid the din of despair, saying that now is precisely the time to seize the initiative and launch the "global revolution" the IEA is calling for. And not just because it will stave off disasters two or three decades away, but also because it can provide the impetus to pull the global economy out of the slump it's in now and put it on a more solid foundation than it's had in at least a generation. British Prime Minister Gordon Brown and French President Nicolas Sarkozy have already taken up the cause. United Nations Secretary-General Ban Ki-moon last month called for a "Green New Deal" that would rebuild and reshape the economy of planet Earth in ways reminiscent of the programs that President Franklin Roosevelt used to revitalize the economy of the United States during the Great Depression. Indeed, even as the slowing economy and falling oil prices make it harder to justify huge new investments in a green economy, there's a strong counterargument that now is precisely the time to make them.

It took a great war, and all the military industries that fed the carnage, to bring America out of the Depression. But to a surprising degree, the world economy has been riding the strength of its hottest sectors ever since. By the 1990s, it was the rise of the Internet, which collapsed with the dotcom bubble and gave way to the housing boom and the financing that paid for it. In each of these recent cases it was the market that discovered and promoted a new engine for growth—creating millions of jobs and trillions in profits worldwide. Between 1996 and 2000, the tech sector created 1.6 million new U.S. jobs, according to Moody's Economy.com—roughly 14 percent of new U.S. job growth. In this decade, the financial sector accounted for the lion's share of U.S. corporate profit, while housing accounted for a staggering 40 percent of new U.S. job growth. Now those two stalled drivers are leading producers of unemployment: Goldman Sachs, for instance, announced a 10 percent staff cut last month.

The world, simply put, needs a new economic driver. Proponents of a Green New Deal argue that massive public investments can lay the groundwork for the private sector to develop whole new industries and create millions of jobs in the near term—and oh, by the way, save the planet in the medium term. "You are not just putting money into hot paper or into a financial-services sector that destroys itself," says Oliver Schäfer, policy director of the European Renewable Energy Council. "You are investing in clean technology, which is real business."

Indeed, in 2008 the promise of jobs may be a stronger incentive to go green than the threat of ice caps melting and coastal cities drowning in 2018 or 2048. In the euro zone, for instance, unemployment is expected to rise from 11.3 million to 14.5 million by the end of next year, pushing the rate up from 7.5 to 9 percent. In the U.S. the rate is 6.1 percent, but is expected to push toward 8 percent by the end of 2009, the highest in 25 years.

Even as a new administration sets to work on that problem, a few European countries have taken the lead pushing forward with substantive green initiatives. According to a recent United Nations report, Germany's $240 billion renewable-energy industry already employs 250,000 people, and by 2020 it is expected to provide more jobs than the country's auto industry. Britain plans to spend $100 billion on 7,000 wind turbines by 2020, and the government claims that will create 160,000 jobs. "I know that some people may be saying that the difficult financial circumstances that the world now faces mean that climate change should move to the back burner of international concern," British Prime Minister Gordon Brown recently said. "I believe the opposite is the case."

> ## "Some people say the financial crisis means that climate change should move to the back burner, but I believe the opposite is the case."
>
> —Gordon Brown

But just how plausible are such plans? Even if they create jobs, will those jobs really contribute to a system that slows or stops global warming? Fatih Birol, the chief economist at the IEA, has overseen the studies calling for a global revolution in the way energy is supplied and used. But he remains pessimistic. He cites a fatal dynamic that is perfectly straightforward. In a recession, consumption of just about all commodities goes down, but so do energy prices—and that discourages the development of alternatives. Nuclear-power plants, vast solar-collection farms, forests of wind turbines, ethanol production, R&D for electric or hydrogen-powered cars and the infrastructure to support them—all require enormous quantities of capital awaiting a fairly distant payoff. When capital and credit are tight, and oil prices suddenly drop (they are less than half what they were in July), private investors are less likely to put billions into a distant clean-energy future. Alternative programs for renewable sources of energy that might make business sense when oil is at $140 a barrel make less sense when it's at $70—and none at all if it drops below $40.

If there is good news, in Birol's view, it's that after the epic interventions in the financial markets over the past few weeks, the notion that the state might intervene massively to redirect the energy market no longer seems extreme, even to the normally laissez-faire British and Americans. Once you open the floodgates of government funding for the banks, why not for green industry, too?

It's the French who offer perhaps the most detailed blueprint for the moment. The particular advantage that Gauls have is that *dirigisme* (state planning) has never been a dirty word in Paris. Massive public investment in rebuilding the economy is what gave the French what they still call *les trente glorieuses,* the 30 glorious years of phenomenal growth after World War II. That was when they made the expensive but prescient decision to build the nuclear-power plants that now supply 80 percent of their electricity with no direct emission of greenhouse gases. So, too, the French *dirigiste* decision to crisscross the country with capital-intensive but energy-efficient high-speed train lines.

Although France was seen as an environmental laggard in the '80s and '90s, when green causes seemed more about lifestyle than survival, over the past year the problems have been defined and addressed with stunning celerity. In October 2007, Sarkozy kept his campaign promise to convene all branches of government, unions, the private sector and other interested parties in a conference similar to the one on the Rue de Grenelle in Paris that ended the quasi revolution of 1968. This Environmental Grenelle, as it's now called, came up with 268 recommendations, many of which have been passed by the parliament. And those have provided Sarkozy with the specifics needed more than ever in the current crisis.

The clear priorities in the new legislation are on practical programs that have an immediate effect on, yes, the job market. First on the list is the construction business: an estimated 25 percent of the country's greenhouse-gas emissions come from energy consumption in buildings. "We're trying to have a 40 percent drop . . . by 2020," says Nathalie Kosciusko-Morizet, the state secretary for ecology, who says the move to make homes, offices and especially public housing better insulated and more energy-efficient will generate some 200,000 of the 500,000 jobs the Grenelle initiatives are supposed to create in France over the next dozen years.

Transportation is another sector that's already been addressed creatively in France. A system that went into effect on Jan. 1 offers a financial bonus for the purchase of cars with low emissions, while there is a tax disincentive (called a *malus*) against buying a car with high emissions. Although the French already drive automobiles that are far more fuel-efficient than most U.S. cars, the move has further transformed the country's taste in automobiles. The sale of pollution-prone used cars has dropped off, while the number of new cars sold in France by Renault, for instance, was up 8.4 percent in September.

> ## "This won't be about sacrificing the future for the present, but putting our country in the best possible situation to face the future."
>
> —Nicolas Sarkozy

Success doesn't come cheap. The bonuses will cost the government up to $265 million this year. Those costs, coming on top of the financial-sector bailouts, will exacerbate France's budget deficit. But Paris says the new expenditures are betting on future energy savings as well as the "formidable follow-on effect" of raising employment and creating dynamic new sectors in the economy. "This time it won't be about sacrificing the future for the present," said Sarkozy, "but on the contrary, putting our country in the best possible situation to face the future."

Can the rest of the world be persuaded to take even more dramatic steps? What of countries like Poland, which produces 94 percent of its electricity from coal? Or China, which pumps more carbon dioxide into the air in eight months than the European Union is likely to save in the next 12 years with all its programs to reduce emissions 20 percent by 2020? Complicated schemes to price carbon emissions and trade carbon credits, some of which are in place, may provide a useful mechanism. So might the costly and almost entirely untested schemes to capture and store the carbon dioxide produced by power plants and factories—the "clean coal" that both McCain and Obama talked about frequently on the campaign trail.

But the political and financial reality is that no government will be so moved by the dire predictions of the world's scientists and the doomsday scenarios on computer models that it will allocate trillions of dollars just to meet those postulated challenges. What governments might do, and some certainly will do, however, is spend huge sums soon to kick-start their economies and create millions of jobs. "The nation is asking for action, and action now," said Franklin Roosevelt when he took office in 1933 and launched the New Deal. Today the global economy—the planet itself—is asking for the same thing.

With William Underhill in London and Jessica Ramirez in Washington, D.C.

Study Finds Humans' Effect on Oceans Comprehensive

Juliet Eilperin

Human activities are affecting every square mile of the world's oceans, according to a study by a team of American, British and Canadian researchers who mapped the severity of the effects from pole to pole.

The analysis of 17 global data sets, led by Benjamin S. Halpern of the National Center for Ecological Analysis and Synthesis in Santa Barbara, Calif., details how humans are reshaping the seas through overfishing, air and water pollution, commercial shipping and other activities. The study, published online yesterday by the journal Science, examines those effects on nearly two dozen marine ecosystems, including coral reefs and continental shelves.

"For the first time we can see where some of the most threatened marine ecosystems are and what might be degrading them," Elizabeth Selig, a doctoral candidate at the University of North Carolina at Chapel Hill and a co-author, said in a statement. "This information enables us to tailor strategies and set priorities for ecosystem management. And it shows that while local efforts are important, we also need to be thinking about global solutions."

The team of scientists analyzed factors that included warming ocean temperatures because of greenhouse gas emissions, nutrient runoff and fishing. They found that the areas under the most stress are "the North and Norwegian seas, South and East China seas, Eastern Caribbean, North American eastern seaboard, Mediterranean, Persian Gulf, Bering Sea, and the waters around Sri Lanka."

Some marine ecosystems are under acute pressure, the scientists concluded, including sea mounts, mangrove swamps, sea grass and coral reefs. Almost half of all coral reefs, they wrote, "experience medium high to very high impact" from humans.

Overall, rising ocean temperatures represent the biggest threat to marine ecosystems.

Pew Environment Group Managing Director Joshua Reichert, whose advocacy organization has launched a campaign to preserve several of the oceans' most ecologically rich regions by creating three to five marine reserves over the next five years, said the study demonstrates that human activity has already transformed "what had been viewed as the Earth's last great bastion of nature."

Reichert added that while it made sense that coastal areas close to dense populations had suffered the most, the scientists' most significant finding was that human effects are reaching even isolated regions.

"As the result of more sophisticated technology and fishing gear that's been able to reach farther underneath the surface . . . we're reaching the remote areas of the sea," Reichert said. "They were off bounds. They're not anymore."

One of the unusual aspects of the new map is its geographic precision. Selig worked with John Bruno, a University of North Carolina marine sciences professor, and Kenneth Casey, a researcher at the National Oceanic and Atmospheric Administration, to create a grid of local ocean temperature variation in which each block measures just 1.5 square miles. Previous data sets spanned areas of nearly 20 square miles.

Polar Distress

With the clock running out in January, the Bush administration, ignoring the concerns of its own scientists and possibly breaking federal law, looks to open a vital stretch of Arctic habitat to offshore oil and gas drilling. So much for saving endangered bears.

DANIEL GLICK

Early each spring pregnant bowhead whales, insulated by up to a foot and a half of fat that helps them withstand frigid Arctic waters, pass north through the Bering Strait and pause in their annual, 3,500-mile migration to give birth to one-ton calves. Following cracks in the melting ice pack off the north coast of Alaska, these bowheads feed on the riot of tiny marine invertebrates that erupts as the returning sunlight sparks a reawakening of an elaborate food chain.

An evolutionary parade swirls alongside the ice leads, from the tiny zooplankton to the baleen whales that miraculously turn krill into life-sustaining blubber. Walrus herds, sometimes numbering in the thousands, use the same moving ice edge like smart shoppers at a sale. Diving from ice platforms, the walruses seek a trove of clams and mussels in the shallow waters below, hauling themselves back onto the ice to rest between feeding forays. Ice-dependent seals—ringed seals, bearded seals, spotted seals, ribbon seals—also trail this floating ice ark, devouring Arctic cod that feed on small shrimplike copepods that dine on phytoplankton.

Almost like a cartoon version of the food chain, with one small fish being eaten by a bigger fish, then that by an even bigger one, the seals in turn fall prey to the lurking top predator cruising this same floating Arctic carnival: the polar bear.

These species and many others, including red-throated and yellow-billed loons and waterfowl including king and spectacled eiders, migrate through or stop to feed or molt in and around a watery place on the globe that cartographers know as the Chukchi Sea. Situated north and west of Point Barrow, the northernmost tip of the United States, the Chukchi is also a focal point for yet another conservation battle along the Alaskan frontier. A recent Bush administration decision to open up offshore oil and gas leases in the Chukchi Sea repeats a now-familiar tilt toward accelerating domestic oil and gas production on public land, no matter what the ecological costs may be.

But this time, by opening up a marine area the size of Pennsylvania to energy development with a controversial lease sale this past February, the Bush administration likely broke federal law, ignored Native Alaskan concerns, censored and disregarded the government's own scientists, and mobilized a national coalition from the conservation community to mount a legal challenge that will be resolved in the courts or Congress—or possibly by the next President. Representative Edward Markey (D-MA), chairman of the House Select Committee on Energy Independence and Global Warming, who unsuccessfully tried to pass legislation to delay the sale, called the Bush administration's approach "regulatory lunacy and a blatant disregard for moral responsibility."

The Chukchi leases appear to be part of a final, concerted push by the Bush administration to open up as much public land as possible to energy development before leaving office. The timing of the lease was especially suspect, since it occurred immediately after the administration postponed a decision about whether to list the polar bear as a threatened species under the Endangered Species Act—and before a thorough environmental review that would include the effects of development on an area that's home to half the U.S. polar bear population. "They've got the clock running and their eye on January," says Stan Senner, executive director of Alaska Audubon. "They're going to squeeze in everything they possibly can, and there's no pretense of balance."

To marine biologists, the Chukchi Sea supports a remarkably rich concentration of offshore Arctic biodiversity. This region is a relatively shallow stretch of water that lies between the Bering Strait to the south, the Beaufort Sea on the east, and the vast stretches of the Arctic Ocean to the north, where waters quickly become too deep and cold to support much life. The Chukchi is, in effect, a last-chance sanctuary, where many of the planet's most charismatic marine mammals, including polar bears and walruses, spend part of their annual cycle of foraging, giving birth, and storing up enough calories to survive and reproduce. In spring, common and thick-billed murres funnel north through the Bering Strait into the Chukchi Sea to reach cliffside nests in such places as Cape Lisburne, from which they take advantage of the tremendous burst of Arctic marine productivity once the high seas are free of ice for the season. "This is the most biologically productive area in the Arctic Ocean," says Lee Cooper, a researcher currently at the University of Maryland's Center for Environmental Science, who has spent more than 20 years studying the Chukchi.

"This is the most biologically productive area in the Arctic Ocean," says Lee Cooper, a researcher currently at the University of Maryland's Center for Environmental Science, who has spent more than 20 years studying the Chukchi.

This region, like much of the circumpolar Arctic, has been undergoing profound ecological changes in recent years, as global warming has affected the Arctic much more dramatically than it has temperate latitudes. Researchers have already documented some of the resulting ecological ripples and fear that bringing industry in will compound the stress of ice-dependent species already coping with retreating summer ice cover, which in 2007 reached its lowest levels since satellite observations began in 1979, according to the National Snow and Ice Data Center.

Environmentalists have waged a two-decades-long campaign to stop oil drilling in Alaska's Arctic National Wildlife Refuge, an onshore area to the east of the Chukchi. But some biologists believe the Chukchi leases are far more problematic, since offshore drilling presents greater challenges and dangers than stationary rigs on land. Lori Quakenbush, the Arctic marine mammals program leader for the Alaska Department of Fish and Game, says that with offshore drilling "there are a lot more risks involved," including the potential for catastrophic oil spills, the harassment of animals through seismic exploration and floating industrial activities, the increase in boat traffic along the same small areas of open water (called leads) that many animals use, and the impacts on Inupiaq subsistence hunters if animals move farther offshore. In the Chukchi, she says, "there are way more species and way more numbers of each species" than in the Arctic National Wildlife Refuge, elsewhere along the Arctic coastal plain, or even in the neighboring Beaufort Sea. "That ramps up the concern."

The Chukchi leasing decision embodies what is possibly the ultimate environmental irony of our time: It is a fight to protect a region already affected by global warming from more energy development, which will only exacerbate those effects once that oil and gas is drilled and burned. With its decision to expedite the contentious Chukchi lease, the Bush administration chose to quite literally add more fuel to the global warming fire.

For the federal Minerals Management Service (MMS), which oversees offshore leases, the biological trove of cetaceans and crustaceans, pinnipeds and ursidae off Alaska's north coast is simply known as Chukchi Sea Sale 193.

Last February 6, at the Z.J. Loussac Public Library in Anchorage, representatives from Shell, ConocoPhillips, and other energy companies packed a meeting room to roll some really big dice in a high-stakes game for the sale. Since this was the first lease sale in the Chukchi in 17 years (those previous leases were relinquished or expired without any oil or gas production), excitement ran high. Armed with proprietary information gleaned from floating seismic sounders they had deployed to survey the ocean floor for telltale signs of hydrocarbons, company representatives submitted sealed bids with potentially several hundred billion dollars of profits at stake.

For years oil companies had shied away from seeking permission to drill in Alaska's offshore areas for several reasons, including the expense of operating in such a challenging environment and the relatively low price of oil. But as predictions of an ice-free Arctic summer moved toward becoming a reality and oil first topped $100 a barrel the month before the sale, companies like Shell took another look and liked what they saw, especially in the lame-duck days of an administration led by oilmen George Bush and Dick Cheney. Leaving little to chance, Shell, a leading player in the Chukchi oil sale, hired a phalanx of former Bush administration officials as well as some Inupiaq leaders to help pave the way, including Paul Stang, former MMS regional supervisor for leasing and environment; Camden Toohey, former special assistant to the Interior Secretary for Alaska Affairs; and George Ahmaogak, former North Slope Borough (NSB) mayor. (The NSB encompasses eight Inupiaq communities.)

Presiding over the sale, Randall Luthi, director of the MMS, opened the auction with an edge of anticipation—not just from the energy companies set to bid tens of millions of dollars for hundreds of nine-square-mile sections but also for protestors gathered to express their opposition. In preparing for the sale, Luthi stated the Bush administration creed that drilling for more domestic oil would reduce the country's reliance on foreign oil at a time when worldwide demand is growing and America's energy use is increasing. He insisted that environmental safeguards would be in place, and noted that some areas close to shore had been placed off-limits. What did go on the auction block, Luthi said, was "one of the last energy frontier areas in North America."

When the day was through, companies had submitted sealed bids totaling $2.6 billion for the rights to oil and gas under nearly 30 million acres of seabed more than 50 miles from the Alaska coast along the outer continental shelf. If any oil or gas were to be extracted, the federal government would receive 12.5 percent royalties for what the MMS has estimated might be as much as 15 billion barrels of oil and 77 trillion cubic feet of natural gas. The state of Alaska would receive virtually no royalty payments, nor would the Alaska Native corporations.

Milling outside the library in minus-12-degree weather, members of the Point Hope community, which had joined a lawsuit contesting the lease, and other Inupiaq protestors held placards that read "Not in My Garden" and "Oil and Polar Bears Don't Mix." Steve Oomittuk, mayor of Point Hope City, said that his ancestors "have hunted and depended on the animals that migrate through the Chukchi Sea for thousands of years."

Even in the Internet age, he said in a statement, Inupiaq communities still depend on subsistence hunting. "This is our garden, our identity, our livelihood. Without it we would not be who we are today."

Inupiaq opposition to the sale hardly came from a knee-jerk, anti-oil position. In fact, their communities have mostly been supportive of onshore oil development, in part because those oil revenues flow into state trusts and Native corporation coffers. Those funds are used to build and sustain infrastructure and social benefits, such as schools, health clinics, university scholarships, and dividends for shareholders in Alaska's Native corporations. On top of that, Inupiaq leaders have long been suspicious of environmental activists from Anchorage or the Lower 48, in part because of the "Save the Whales" campaigns that once threatened their subsistence hunt.

But the current play to extend onshore energy development into the ocean prompted much of the Inupiaq community to draw a line on the shore. For each of the Native communities along the North Slope—Point Hope, Wainwright, Point Lay, Barrow, Nuiqsut and Kaktovik—the spring and fall bowhead whale hunt remains a cultural cornerstone as well as an ongoing key to survival for many who rely on a seasonal catch of fish, seal, walrus, and whale meat for cultural and alimentary sustenance.

The rhythms of arctic life—both human and non-human—thrum in concert with the seasons. In February, shortly after the sun rises above the horizon in most North Slope communities, Inupiaq whalers begin preparations for the spring bowhead hunt, readying their traditional bearded seal- or walrus-skin boats, called *umiak*. Women sew boat covers of seal or walrus hide onto the wooden frames using thread made from dried caribou sinew, waterproofing their stitches by rubbing them with seal or whale oil and painting them white so they blend in with the ice. The men on the 50 or so whaling crews from a dozen communities that have traditional rights to catch bowheads prepare their harpoons and camping gear. By April the spring hunt usually begins. After the hunt, these communities celebrate their success with blanket tosses and traditional feasts while whaling captains divvy up the *muktuk*, that same life-sustaining whale blubber, and meat among the villagers.

Inupiaq resistance to the Chukchi drilling leases centers on the threats to an ancient whaling culture. On January 1 the Native village of Point Hope (the westernmost of the seven coastal Inupiaq villages in the NSB), the city of Point Hope, the Inupiaq community of Arctic Slope, and the Resisting Environmental Destruction on Indigenous Lands (REDOIL) network joined with national conservation organizations in a legal challenge to the leases. The lawsuit, which was filed by 13 plaintiffs, including Audubon, the Sierra Club, and the Wilderness Society, claims that the expedited environmental review process minimized the potential environmental consequences of oil and gas spills and used outdated or insufficient data. The suit also charges that the process ignored existing data suggesting that animals like the spectacled eider, which is listed as threatened under the Endangered Species Act, would suffer from offshore drilling.

In December 2006, during that review, Edward A. Itta, the current NSB mayor, wrote an unusually strong letter to the MMS. It concluded that oil and gas leasing was simply too problematic to support, given the lack of good environmental data, the dangers of drilling remote areas in harsh conditions, and the "failure" of the industry to show it could respond to oil spills. "It remains our strong belief that oil and gas leasing, exploration, and development should not occur in the Chukchi Sea," Itta argued.

His concerns reflect a widespread sentiment among scientific researchers like Quakenbush, the Alaska Fish and Game Department marine mammals expert, who insists that it's virtually impossible to gauge potential impacts when there is very little research to draw on—and when the region is changing so fast. She and others have been using satellite telemetry to track the movements of bowhead whales and other species. Their data indicate that the Chukchi is chock-full of animals—bowheads, beluga whales, polar bears, walruses, and various migratory bird species—that migrate through during the spring and fall.

Data that does exist about the long-term ecological consequences of oil spills, like $400 million worth of research completed in the aftermath of the *Exxon Valdez* spill in 1989, was virtually disregarded, says University of Alaska professor and marine conservation specialist Rick Steiner. So was research regarding the threat of invasive species carried in on construction and operations traffic. "Anything that would challenge the outcome or slow it down was marginalized or completely ignored," he says. Leaked e-mails from agency scientists involved with the Chukchi leases confirmed a pattern of intimidation, secrecy, and suppression of information that might have delayed or stopped the leases, according to Jeff Ruch, executive director of Public Employees for Environmental Responsibility, a Washington, D.C.–based group that defends federal whistleblowers.

The MMS's own public estimates for potential calamity ought to be sounding a loud alarm. In the final environmental impact statement (EIS) that governed the Chukchi lease sale, the government forecast a 33 percent to 51 percent chance of a large spill in the Chukchi. (By contrast, says Steiner, the government had predicted that an *Exxon Valdez*–type spill would occur once every 241 years; it happened the 12th year after tankers started sailing from Valdez.) "If a large spill were to occur, the analysis identifies potentially significant impacts to bowhead whales, polar bears, essential fish habitat, marine and coastal birds, subsistence hunting, and archaeological sites," said the EIS.

The MMS also acknowledges that walruses, a species that is notoriously sensitive and has been known to stampede one another to death when disturbed, would be harmed by development. Furthermore, bowhead whales are susceptible to damage from seismic activity. Threatened eider species could face potentially significant mortality. Oil spills, in particular, would cause a slew of devastating effects that would ripple through the food chain. In the final EIS, wrote the MMS, "There is a high potential for marine and coastal birds to experience disturbance and habitat alteration."

Even the research that the MMS quoted is being mischaracterized, says the University of Maryland's Cooper. While researching a recent application for a National Science Foundation grant to study the Chukchi, Cooper reviewed the MMS materials only to discover that the federal agency was using outdated information and had glossed over recent research, like that done by Cooper and his colleagues. For example, the scientists had discovered a "submarine canyon" in the Chukchi that is a forest of corals, tunicates (filter feeders like sea squirts), and vitally important benthic marine life. "They're working with very old information," says Cooper. "If people knew what was on the bottom there, they would fence it out."

Nevertheless, the agency's conclusion was one of the federal government's most unfortunate acronyms: FONSI, or "finding of no significant impact."

Offshore oil activity, even in relatively benign environments, has resulted in 117 spills in outer continental shelf waters since 2000 alone, according to MMS data. Offshore oil platforms have a checkered history even in more moderate seas below the Arctic Circle. Last December Statoil-Hydro, a Norwegian company that won some of the Chukchi leases, announced that 25,000 barrels of oil had spilled at one of its North Sea oilfields. The standard practices of reacting to oil spills at sea by burning them, by spreading dispersants, or by mechanical recovery would be nearly impossible in the icy conditions that still exist during most of the Arctic year. "They don't even begin to have the technology to clean up oil in an environment like the Chukchi Sea," says Alaska Audubon's Senner.

Scientists inside and outside federal agencies warn that increasing the floating human footprint from offshore rigs, seismic exploration, and more transport vessels may well push a reeling environment over the edge. Inserting tool pushers, roughnecks, roustabouts, mud engineers, derrick hands, geologists, and other oil platform workers into a world of stressed-out seals, worried walruses, and imperiled polar bears strikes many wildlife biologists and oceanographers as a singularly bad idea. Add to that the likelihood of oil spills in an icy, stormy clime, and ecologists warn of disasters like the *Exxon Valdez* spill—or worse.

Today the heralds of climate change have been awarded the Nobel Prize. Species are disappearing at a horrifying rate. And the United States' leadership in environmental stewardship has been sorely tainted. These events in the "polar bear seas" thus raise concerns across the country.

Energy policy—especially support for renewable energy and greater conservation—will certainly become a 2008 campaign issue. Experts in and out of industry acknowledge that the U.S. appetite for fossil fuels will never be sated by domestic production. As the planet warms and the unintended consequences of burning fossil fuels become more apparent, it is likely that the cries to protect some of this country's last undeveloped lands will only grow louder.

In the eyes of University of Alaska professor Steiner, however, the Chukchi sales represent "a worrisome new phase in our addiction for oil." The dire need for the world to combat global warming by weaning itself from hydrocarbons is bound to bring up analogies to reckless junkies. Says Steiner, "We are at the desperate stage where we are accepting more and more risk to get the next fix."

DANIEL GLICK, a frequent contributor to *Audubon,* is a Colorado-based writer and the author of *Monkey Dancing: A Father, Two Kids, and a Journey to the Ends of the Earth* (PublicAffairs).

What China Can Learn from Japan on Cleaning up the Environment

Japan's experience provides an excellent road map for anyone who wants to gauge how China can deal with its environmental problems.

BILL EMMOTT

Analyses of China veer wildly between the awe-struck and the apocalyptic, between the view that the country has achieved wholly exceptional and unstoppable economic growth and that its growth must end in disaster, probably soon. Lately, the polluted environment has taken pride of place in the apocalyptic argument, on the theory that China's current economic model is unsustainable, whether because of the domestic implications of dirty growth, the planetary implications, or both. The worldwide focus on the country during the recent Beijing Olympic Games gave greater prominence than ever to these environmental issues.

But both the awestruck and the apocalyptic analyses miss the blend of helpful precedent and hard-fought adaptation that has guided China's course since market-based liberalization began, in the 1970s. Far from being unprecedented, the broad shape and nature of the country's growth from the 1980s onward has been pretty similar to the pattern shown in earlier decades by Japan, South Korea, Taiwan, and other East Asian success stories. Like them, moreover, China has had to adapt constantly as circumstances changed and new challenges arose to confront policy makers and managers alike.

In the late 1980s, the country was roiled by hyperinflation and the post-Tiananmen reform crisis; in the 1990s, by the mountain of nonperforming bank loans and the huge task of closing or selling off dud state-owned enterprises. Today's vast pollution problem is just the latest in a series of obstacles that have fed the pessimistic tendency in Western commentary, The fact that China succeeded in overcoming past obstacles doesn't guarantee future success, of course. But the good news is that China's current situation is not at all unprecedented. To its east lies a country that provides an excellent road map for policy makers and executives who wish to gauge how China can come to grips with its environmental problems during the next decade. This country is called Japan.

Think Back 40 Years

During the current decade, China's economy has become almost a caricature of the East Asian model pioneered by Japan and South Korea. Investment has surged, taking the share of capital formation in GDP to 45 percent. High and rising domestic savings, accumulated by households, companies, and the government, have made that level of investment easy to finance. The renminbi, undervalued by most measures, has been allowed to appreciate only modestly and only since 2005. An extraordinary $2 trillion in foreign-exchange reserves has been built up to keep the currency cheap, even as the surplus on the current account of China's balance of payments has climbed to more than 9 percent of GDP.

Yet for all the success of that low-cost, cheap-currency, high-investment model, China now faces twin challenges. The first is a sudden swing, in the past two years, from deflation to quite rapid inflation driven not only by oil and food prices but also rising wages, The second is a sharp worsening of air and water pollution, because resource- and emission-intensive heavy industries—such as aluminum, chemicals, and steel—have led the country's economic growth. Until 2002, China's energy intensity (energy used per unit of GDP) had been declining for two decades. Since then, it has been rising, along with emissions of greenhouse gases. In 2007, China overtook the United States as the world's biggest producer (in absolute terms, rather than emissions per head).

It looks daunting—and it is. But the twin challenges are also strikingly similar to those that beset Japan's economy as it entered the 1970s. Like China now and South Korea in the 1980s, Japan, in its high-growth years of the 1950s and '60s, propelled its economy through investment, which rose to 40 percent of GDP in 1970. Under the Bretton Woods system, the yen enjoyed a fixed rate against the dollar, and this made the country's currency more and more undervalued as Japanese industry became increasingly competitive. Fixed exchange rates and capital controls prevented Japan from building up a current-account surplus and foreign-exchange reserves as large as China's today. Nonetheless, surpluses caused international friction, especially with the United States. Meanwhile, environmental problems caused domestic friction.

Many will recall that in the early 1970s, Japan encountered two shocks. The "Nixon Shock" came when the US president alarmed Japan by visiting Mao's China and by abruptly abandoning fixed exchange rates, forcing a sudden revaluation of the yen. Then the oil shock of 1973 sharply worsened Japan's terms of trade and sent its inflation rate soaring to 25 percent. What is often forgotten is that Japan, like China today, had a disastrously dirty environment, for its heavy industry was highly polluting and unrestrained by environmental-control laws, Consider this passage from a 1975 book by Frank Gibney, *Japan: The Fragile Superpower*[1]:

> Most of the beautifully scenic Inland Sea was hopelessly dirtied by the so-called red tides of polluted waters from the factories on its shores. Smog warnings became regular and asthma sufferers began trekking to the hospitals. Regional complaints and petitions about pollution, about 20,000 in 1966, had risen to 76,000 in 1971. . . . In the south, hundreds of people fell ill from eating the local fish. Many died. Similar problems occurred in the north, with mercury-filled drainage from one factory and where a painful bone disease was caused by cadmium.

Now compare that passage with one from a book, published last year, about China:

> For two decades, the government treated environmental protection as a distraction from economic growth. . . . Breakneck industrialization produced some of the worst air and water pollution in the world. According to environmental officials, acid rain is falling on one-third of the country, half of tie water in its seven largest rivers is "completely useless" . . . one-third of the urban population is breathing polluted air. . . . More than 70 percent of the rivers and lakes are polluted, and groundwater in 90 percent of the cities is tainted.

That extract comes from a book by Susan L. Shirk called—guess what—*China: Fragile Superpower.*[2] Clearly, fragility and Asian superpowerdom go together in Western minds.

The place in Japan where thousands fell ill from eating fish was Minamata Bay, in Kyushu. There, the Japanese company Chisso poured methylmercury into the sea during the 1950s and 1960s, with disastrous consequences for the health of the local people—consequences that are still seen today. The cadmium poisoning, caused by Mitsui Mining & Smelting, produced a disease known as *itai-itai*, or "it hurts, it hurts." The mercury-filled drainage came from the company Showa Denko, whose effluent triggered a second outbreak of Minamata disease, in Niigata Prefecture, in northwest Japan.

If you look at a photograph taken in a Japanese city in 1970—Tokyo, Osaka, or Kitakyushu, or any city with industry right at its heart—you will find a strong resemblance to most big Chinese cities today, including Beijing, Guangzhou, Shanghai, and Chongqing. In these photos, the air is thick with smog and dust. The rare days when blue skies could be seen were causes for celebration. If you go to any of those Japanese cities now, however, the air is completely different. Blue skies are no longer a rarity, people need wear face masks only when they suffer from colds, not in an attempt to filter out pollution (as they did during the 1960s), and it is no longer hazardous to be a traffic policeman.

A combination of two things cleaned the air in Japan. One was popular protest, which even in a democracy dominated by a single party, the Liberal Democrats, forced government policy to change. The country's first proper environmental laws were passed in the early 1970s, when its first environment agency was created. The second was macroeconomic: the revaluation of the yen, combined with the oil shock and the ensuing inflation. This sudden change in Japan's circumstances brought about an abrupt switch in its industrial structure. The low-cost model was dead. Capital investment in heavy, polluting industry began to drop. Energy had become more expensive, and new taxes made it more expensive still. Companies adopted energy-saving and more efficient technologies and started to make products, especially cars, suited to the new, cleaner times. Also at this point, electronics companies, encouraged by the government, made big investments in new high-tech gadgetry, which led the economy in a new direction.

Following the Japanese Example

The coincidence of inflation and environmental protests made Japanese industry's sharp move upmarket inescapable. This move was also extremely successful, leading the way toward a further decade and a half of world-beating growth. At first, it was tough for the economy as a whole, but the Japanese government's strong fiscal position enabled it to run growth-supporting deficits for much of the next decade.

Can China do the same? Everything there is on a larger scale than elsewhere: only India comes close in population, and only Russia, Canada, and the United States are comparable in geographic size. This can make China's problems look more daunting than those of other countries, But that is misleading, for alongside the scale of the country's pollution problems must be placed the scale of its resources to deal with them—if it chooses to do so, China has plenty of officials to enforce laws, the world's largest security forces, and a central-government budget that is now happily in surplus.

Its inflationary pressures so far are milder than Japan's in the '70s (an annual rate of 8 percent, versus 25 percent in Japan in 1974–75). Its currency policy is in its own hands rather than Richard Nixon's, and the pressure from environmental protests, while real and growing, is muted. The biggest problem is decentralization: China has excellent environmental laws, but local governments have so far tended to ignore them.

Nevertheless, China's central-government policy makers have already shown, through their speeches, that they too see inflation and the environment as the biggest obstacles the economy must now surmount. The ground has been prepared for changes both in macroeconomic and environmental policy. If they are carried out in both areas, the two sets of changes would reinforce each other. Without a faster revaluation of the renminbi, monetary policy will not come to grips with inflation, as interest rates will have to stay too low. Without greater central-government control over local authorities, environmental laws will not be enforced. Most of all, without powerful market signals that the low-cost, resource- and capital-intensive phase of China's development is over, neither private enterprises nor state-owned ones (which dominate heavy industry and are the biggest polluters) will move upmarket and clean up their processes.

Like Japan in the 1970s, China has the advantage of strong public finances, which can support the economy through its next transition and finance an environmental cleanup. Like Japan in those years as well, China is a country where the state plays a large role, which could influence industrial restructuring and investment in R&D. Unlike Japan, China could benefit from the worldwide concern over climate change: just as in the 1990s membership in the World Trade Organization was used to overcome domestic opponents of reform, so too negotiations over global climate change could now be used to overcome resistance to environmental enforcement. The biggest difficulty is that in China, unlike Japan, neither politics nor policy is tightly centralized—they cannot be, given China's scale and complexity. Beneficial change won't come easily, But it can indeed come.

References

1. Frank Gibney, *Japan: The Fragile Superpower,* New York: W. W. Norton & Co., 1975.

2. Susan L. Shirk, *China: Fragile Superpower,* New York: Oxford University Press, 2007.

BILL EMMOTT is the former editor of *The Economist* and author of the recently published book *Rivals: How the Power Struggle Between China, India and Japan Will Shape Our Next Decade.*

UNIT 3

The Region

Unit Selections

Key Points to Consider

- To what regions do you belong?

- Why are maps and atlases so important in discussing and studying regions?

- What major regions in the world is experiencing change? Which ones seem not to change at all? What are some reasons for the differences?

- What regions in the world are experiencing tensions? What are the reasons behind these tensions? How can the tensions be eased?

- What are the long-term impacts of drought in the western United States?

- Why are regions in Africa suffering so greatly?

- What is the solution to the turmoil in Darfur?

- Will agricultural output keep pace with population growth in the 21st century?

- Why is regional study important?

- Discuss geographical examples of accessibility.

Student Website

www.mhcls.com

Internet References

AS at UVA Yellow Pages: Regional Studies
http://xroads.virginia.edu/~YP/regional.html

Can Cities Save the Future?
http://www.huduser.org/publications/econdev/habitat/prep2.html

Minnesota Department of Transportation
www.dot.state.mn.us

NewsPage
http://www.individual.com

World Regions & Nation States
http://www.worldcapitalforum.com/worregstat.html

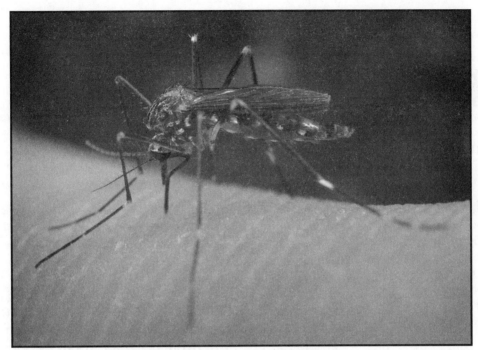

© Courtesy of the Center for Disease Control

The region is one of the most important concepts in geography. The term has special significance for the geographer, and it has been used as a kind of area classification system in the discipline.

Two of the regional types most used in geography are "uniform" and "nodal." A uniform region is one in which a distinct set of features is present. The distinctiveness of the combination of features marks the region as being different from others. These features include climate type, soil type, prominent languages, resource deposits, and virtually any other identifiable phenomenon having a spatial dimension.

The nodal region reflects the zone of influence of a city or other nodal place. Imagine a rural town in which a farm-implement service center is located. Now imagine lines drawn on a map linking this service center with every farm within the area that uses it. Finally, imagine a single line enclosing the entire area in which the individual farms are located. The enclosed area is defined as a nodal region. The nodal region implies interaction. Regions of this type are defined on the basis of banking linkages, newspaper circulation, and telephone traffic, among other things.

This unit presents examples of a number of regional themes. These selections can provide only a hint of the scope and diversity of the region in geography. There is no limit to the number of regions; there are as many as the researcher sets out to define.

The first article highlights "brainpower" in India and the country's rapid economic growth. "Hints of a Comeback . . ." relates the recent resurgence of the Erie Canal. The next article recounts life expectancy declines in the Midwest and the South. A review of China's inroads into African countries follows. An article on the geopolitical importance of the Black Sea region is next.

"Half-way from Rags to Riches . . ." reviews Vietnam's path to becoming a modern industrial country. The last article deals with the persistence of malaria, particularly in the developing world.

The Rise of India

Growth is only just starting, but the country's brainpower is already reshaping Corporate America.

MANJEET KRIPALANI AND PETE ENGARDIO

Pulling into General Electric's John F. Welch Technology Center, a uniformed guard waves you through an iron gate. Once inside, you leave the dusty, traffic-clogged streets of Bangalore and enter a leafy campus of low buildings that gleam in the sun. Bright hallways lined with plants and abstract art—"it encourages creativity," explains a manager—lead through laboratories where physicists, chemists, metallurgists, and computer engineers huddle over gurgling beakers, electron microscopes, and spectrophotometers. Except for the female engineers wearing saris and the soothing Hindi pop music wafting through the open-air dining pavilion, this could be GE's giant research-and-development facility in the upstate New York town of Niskayuna.

It's more like Niskayuna than you might think. The center's 1,800 engineers—a quarter of them have PhDs—are engaged in fundamental research for most of GE's 13 divisions. In one lab, they tweak the aerodynamic designs of turbine-engine blades. In another, they're scrutinizing the molecular structure of materials to be used in DVDs for short-term use in which the movie is automatically erased after a few days. In another, technicians have rigged up a working model of a GE plastics plant in Spain and devised a way to boost output there by 20%. Patents? Engineers here have filed for 95 in the U.S. since the center opened in 2000.

Pretty impressive for a place that just four years ago was a fallow plot of land. Even more impressive, the Bangalore operation has become vital to the future of one of America's biggest, most profitable companies. "The game here really isn't about saving costs but to speed innovation and generate growth for the company," explains Bolivian-born Managing Director Guillermo Wille, one of the center's few non-Indians.

The Welch center is at the vanguard of one of the biggest mind-melds in history. Plenty of Americans know of India's inexpensive software writers and have figured out that the nice clerk who booked their air ticket is in Delhi. But these are just superficial signs of India's capabilities. Quietly but with breathtaking speed, India and its millions of world-class engineering, business, and medical graduates are becoming enmeshed in America's New Economy in ways most of us barely imagine.

"India has always had brilliant, educated people," says tech-trend forecaster Paul Saffo of the Institute for the Future in Menlo Park, Calif. "Now Indians are taking the lead in colonizing cyberspace."

"Just like China drove down costs in manufacturing and Wal-Mart in retail, India will drive down costs in services," says an Indian IT exec.

This techno take-off is wonderful for India—but terrifying for many Americans. In fact, India's emergence is fast turning into the latest Rorschach test on globalization. Many see India's digital workers as bearers of new prosperity to a deserving nation and vital partners of Corporate America. Others see them as shock troops in the final assault on good-paying jobs. Howard Rubin, executive vice-president of Meta Group Inc., a Stamford (Conn.) information-technology consultant, notes that big U.S. companies are shedding 500 to 2,000 IT staffers at a time. "These people won't get reabsorbed into the workforce until they get the right skills," he says. Even Indian execs see the problem. "What happened in manufacturing is happening in services," says Azim H. Premji, chairman of IT supplier Wipro Ltd. "That raises a lot of social issues for the U.S."

No wonder India is at the center of a brewing storm in America, where politicians are starting to view offshore outsourcing as the root of the jobless recovery in tech and services. An outcry in Indiana recently prompted the state to cancel a $15 million IT contract with India's Tata Consulting. The telecom workers' union is up in arms, and Congress is probing whether the security of financial and medical records is at risk. As hiring explodes in India, the jobless rate among U.S. software engineers has more than doubled, to 4.6%, in three years. The rate is 6.7% for electrical engineers and 7.7% for network administrators. In all, the Bureau of Labor Statistics reports that 234,000 IT professionals are unemployed.

Where India Is Making an Impact

Software
India is now a major base for developing new applications for finance, digital appliances, and industrial plants.

IT Consulting
Companies such as Wipro, Infosys, and Tata are managing U.S. IT networks and re-engineering business processes.

Call Centers
Thousands of Indians handle customer service and process insurance claims, loans, bookings, and credit-card bills.

Chip Design
Intel, Texas Instruments, and many U.S. startups use India as an R&D hub for mircroprocessors and multimedia chips.

. . .And Where It's Going Next

Financial Analysis
Research for Wall Street will surge as U.S. investment banks, brokerages, and accounting firms open big offices.

Industrial Engineering
India does vital R&D for GE Medical, GM, engine maker Cummins, Ford, and other manufactures plan big engineering hubs.

Analytics
U.S. companies are hiring Indian math experts to devise models for risk analysis, consumer behavior, and industrial processes.

Drug Research
As U.S. R&D costs soar, India is expected to be a center for biotechnology and clinical testing.

Data: *BusinessWeek*

The biggest cause of job losses, of course, has been the U.S. economic downturn. Still, there's little denying that the offshore shift is a factor. By some estimates, there are more IT engineers in Bangalore (150,000) than in Silicon Valley (120,000). Meta figures at least one-third of new IT development work for big U.S. companies is done overseas, with India the biggest site. And India could start grabbing jobs from other sectors. A.T. Kearney Inc. predicts that 500,000 financial-services jobs will go offshore by 2008. Indiana notwithstanding, U.S. governments are increasingly using India to manage everything from accounting to their food-stamp programs. Even the U.S. Postal Service is taking work there. Auto engineering and drug research could be next.

More Science in Schools

Tech luminary Andrew S. Grove, CEO of Intel Corp., warns that "it's a very valid question" to ask whether America could eventually lose its overwhelming dominance in IT, just as it did in electronics manufacturing. Plunging global telecom costs, lower engineering wages abroad, and new interactive-design software are driving revolutionary change, Grove said at a software conference in October. "From a technical and productivity standpoint, the engineer sitting 6,000 miles away might as well be in the next cubicle and on the local area network." To maintain America's edge, he said, Washington and U.S. industry must double software productivity through more R&D investment and science education.

But there's also a far more positive view—that harnessing Indian brainpower will greatly boost American tech and services leadership by filling a big projected shortfall in skilled labor as baby boomers retire. That's especially possible with smarter U.S. policy. Companies from GE Medical Systems to Cummins to Microsoft to enterprise-software firm PeopleSoft that are hiring

WHY **CORPORATE AMERICA** IS BEATING A PATH TO INDIA

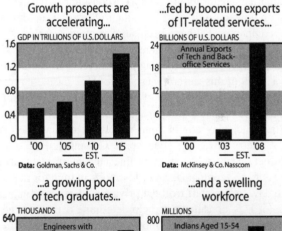

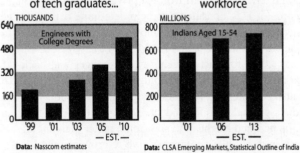

in India say they aren't laying off any U.S. engineers. Instead, by augmenting their U.S. R&D teams with the 260,000 engineers pumped out by Indian schools each year, they can afford to throw many more brains at a task and speed up product launches, develop more prototypes, and upgrade quality. A top electrical or chemical engineering grad from Indian Institutes of Technology (IITS) earns about $10,000 a year—roughly one-eighth of U.S. starting pay. Says Rajat Gupta, an IIT-Delhi grad and senior

Who's Bulking Up

Some of the biggest U.S. players in India

Company	Purpose	India Staff
GE Captial Services	Back-office work	16,000
GE's John Welsh Tech Center	Product R&D	1,800
IBM Global Services	IT services, software	10,000
Oracle	Software, services	6,000**
EDS	IT services	3,500[†]
Texas Instruments	Chip design	900
Intel	Chip design, software	1,700
J.P. Morgan Chase	Back-office, analysis	1,200

*By 2005 **Unspecified [†]By 2004 Data: Company reports, Nasscom, Evalueserve

partner at consulting firm McKinsey & Co.: "Offshoring work will spur innovation, job creation, and dramatic increases in productivity that will be passed on to the consumer."

Whether you regard the trend as disruptive or beneficial, one thing is clear. Corporate America no longer feels it can afford to ignore India. "There's just no place left to squeeze" costs in the U.S., says Chris Disher, a Booz Allen Hamilton Inc. outsourcing specialist. "That's why every CEO is looking at India, and every board is asking about it." neoIT, a consultant advising U.S. clients on how to set up shop in India, says it has been deluged by big companies that have been slow to move offshore. "It is getting to a state where companies are literally desperate," says Bangalore-based neoIT managing partner Avinash Vashistha.

As a result of this shift, few aspects of U.S. business remain untouched. The hidden hands of skilled Indians are present in the interactive websites of companies such as Lehman Brothers and Boeing, display ads in your Yellow Pages, and the electronic circuitry powering your Apple Computer iPod. While Wall Street sleeps, Indian analysts digest the latest financial disclosures of U.S. companies and file reports in time for the next trading day. Indian staff troll the private medical and financial records of U.S. consumers to help determine if they are good risks for insurance policies, mortgages, or credit cards from American Express Co. and J.P. Morgan Chase & Co.

By 2008, forecasts McKinsey, IT services and back-office work in India will swell fivefold, to a $57 billion annual export industry employing 4 million people and accounting for 7% of India's gross domestic product. That growth is inspiring more of the best and brightest to stay home rather than migrate. "We work in world-class companies, we're growing, and it's exciting," says Anandraj Sengupta, 24, an IIT grad and young star at GE's Welch Centre, where he has filed for two patents. "The opportunities exist here in India."

If India can turn into a fast-growth economy, it will be the first developing nation that used its brainpower, not natural resources or the raw muscle of factory labor, as the catalyst. And this huge country desperately needs China-style growth. For all its R&D labs, India remains visibly Third World. IT service exports employ less than 1% of the workforce. Per-capita income

is just $460, and 300 million Indians subsist on $1 a day or less. Lethargic courts can take 20 years to resolve contract disputes. And what pass for highways in Bombay are choked, crumbling roads lined with slums, garbage heaps, and homeless migrants sleeping on bare pavement. More than a third of India's 1 billion citizens are illiterate, and just 60% of homes have electricity. Most bureaucracies are bloated, corrupt, and dysfunctional. The government's 10% budget deficit is alarming. Tensions between Hindus and Muslims always seem poised to explode, and the risk of war with nuclear-armed Pakistan is ever-present.

So it's little wonder that, compared to China with its modern infrastructure and disciplined workforce, India is far behind in exports and as a magnet for foreign investment. While China began reforming in 1979, India only started to emerge from self-imposed economic isolation after a harrowing financial crisis in 1991. China has seen annual growth often exceeding 10%, far better than India's decade-long average of 6%.

In the Valley's Marrow

Still, this deep source of low-cost, high-IQ, English-speaking brainpower may soon have a more far-reaching impact on the U.S. than China. Manufacturing—China's strength—accounts for just 14% of U.S. output and 11% of jobs. India's forte is services—which make up 60% of the U.S. economy and employ two-thirds of its workers. And Indian knowledge workers are making their way up the New Economy food chain, mastering tasks requiring analysis, marketing acumen, and creativity.

This means India is penetrating America's economic core. The 900 engineers at Texas Instruments Inc.'s Bangalore chip-design operation boast 225 patents. Intel Inc.'s Bangalore campus is leading worldwide research for the company's 32-bit microprocessors for servers and wireless chips. "These are corporate crown jewels," says Intel India President Ketan Sampat. India is even getting hard-wired into Silicon Valley. Venture capitalists say anywhere from one-third to three-quarters of the software, chip, and e-commerce startups they now back have Indian R&D teams from the get-go. "We can barely imagine investing in a company without at least asking what their plans are for India," says Sequoia Capital partner Michael Moritz, who nurtured Google, Flextronics, and Agile Software. "India has seeped into the marrow of the Valley [see box]."

It's seeping into the marrow of Main Street. This year, the tax returns of some 20,000 Americans were prepared by $500-a-month CPAs such as Sandhya Iyer, 24, in the Bombay office of Bangalore's MphasiS. After reading scanned seed and fertilizer invoices, soybean sales receipts, W2 forms, and investment records from a farmer in Kansas, Iyer fills in the farmer's 82-page return. "He needs to amortize these," she types next to an entry for new machinery and a barn. A U.S. CPA reviews and signs the finished return. Next year, up to 200,000 U.S. returns will be done in India, says CCH Inc. in Riverwoods, Ill., a supplier of accounting software. And it's not only Big Four firms that are outsourcing. "We are seeing lots of firms with 30 to 200 CPAs—even single practitioners," says CCH Sales Vice-President Mike Sabbatis.

A top electrical-engineering grad from one of the six Indian Institutes of Technology fetches about $10,000 a year.

The gains in efficiency could be tremendous. Indeed, India is accelerating a sweeping reengineering of Corporate America. Companies are shifting bill payment, human resources, and other functions to new, paperless centers in India. To be sure, many corporations have run into myriad headaches, ranging from poor communications to inconsistent quality. Dell Inc. recently said it is moving computer support for corporate clients back to the U.S. Still, a raft of studies by Deloitte Research, Gartner, Booz Allen, and other consultants find that companies shifting work to India have cut costs by 40% to 60%. Companies can offer customer support and use pricey computer gear 24/7. U.S. banks can process mortgage applications in three hours rather than three days. Predicts Nandan M. Nilekani, managing director of Bangalore-based Infosys Technologies Ltd.: "Just like China drove down costs in manufacturing and Wal-Mart in retail," he says, "India will drive down costs in services."

GE Capital saves up to $340 million a year by performing some 700 tasks in India.

But deflation will also mean plenty of short-term pain for U.S. companies and workers who never imagined they'd face foreign rivals. Consider America's $240 billion IT-services industry. Indian players led by Infosys, Tata, and Wipro got their big breaks during the Y2K scare, when U.S. outfits needed all the software help they could get. Indians still have less than 3% of the market. But by undercutting giants such as Accenture, IBM, and Electronic Data Systems by a third or more for software and consulting, they've altered the industry's pricing. "The Indian labor card is unbeatable," says Chief Technology Officer John

Parkinson of consultant Cap Gemini Ernst & Young. "We don't know how to use technology to make up the difference."

Wrenching Change

Many U.S. white-collar workers are also in for wrenching change. A study by Mckinsey global institute, which believes offshore outsourcing is good, also notes that only 36% of Americans displaced in the previous two decades found jobs at the same or higher pay. The incomes of a quarter of them dropped 30% or more. Given the higher demands of employers, who want technicians adept at innovation and management, it could take years before today's IT workers land solidly on their feet.

India's IT workers, in contrast, sense an enormous opportunity. The country has long possessed some basics of a strong market-driven economy: private corporations, democratic government, Western accounting standards, an active stock market, widespread English use, and schools strong in computer science and math. But its bureaucracy suffocated industry with onerous controls and taxes, and the best scientific and business minds went to the U.S., where the 1.8 million Indian expatriates rank among the most successful immigrant groups.

Now, many talented Indians feel a sense of optimism India hasn't experienced in decades. "IT is driving India's boom, and we in the younger generation can really deliver the country from poverty," says Rhythm Tyagi, 22, a master's degree student at the new Indian Institute of Information Technology in Bangalore. The campus is completely wired for Wi-Fi and boasts classrooms with videoconferencing to beam sessions to 300 other colleges.

That confidence is finally spurring the government to tackle many of the problems that have plagued India for so long. Since 2001, Delhi has been furiously building a network of highways. Modern airports are next. Deregulation of the power sector should lead to new capacity. Free education for girls to age 14 is a national priority. "One by one, the government is solving the bottlenecks," says Deepak Parekh, a financier who heads the quasi-governmental Infrastructure Development Finance Co.

Future Vision

India also is working to assure that it will be able to meet future demand for knowledge workers at home and abroad. India produces 3.1 million college graduates a year, but that's expected to double by 2010. The number of engineering colleges is slated to grow 50%, to nearly 1,600, in four years. Of course, not all are good enough to produce the world-class grads of elite schools like the IITs, which accepted just 3,500 of 178,000 applicants last year. So there's a growing movement to boost faculty salaries and reach more students nationwide through broadcasts. India's rich diaspora population is chipping in, too. Prominent Indian Americans helped found the new Indian School of Business, a tie-up with Wharton School and Northwestern University's Kellogg Graduate School of Management that lured most of its faculty from the U.S. Meanwhile, the six IIT campuses are tapping alumni for donations and research links with Stanford, Purdue, and other top science universities. "Our mission is to

Where China Is Way Ahead. . .

GROWTH GDP has risen an average of 8% for the past decade, compared with India's 6%.

INFRASTRUCTURE Highways, ports, power sector, and industrial parks are far superior.

FOREIGN INVESTMENT China lures $50 billion-plus a year. India gets $4 billion.

EXPORTS $266 billion reported in 2002 was more than four times India's total.

. . .Where India Has the Edge

LANGUAGE English gives India a big edge in IT services and back-office work.

CAPITAL MARKETS Private firms have readier access to funding. China favors state sector.

LEGAL SYSTEM Contract law and copyright protection are more developed than in China.

DEMOGRAPHICS Some 53% of India's population is under age 25, vs. 45% in China.

become one of the leading science institutions in the world," says director Ashok Mishra of IIT-Bombay, which has raised $16 million from alumni in the past five years.

If India manages growth well, its huge population could prove an asset. By 2020, 47% of Indians will be between 15 and 59, compared with 35% now. The working-age populations of the U.S. and China are projected to shrink. So India is destined to have the world's largest population of workers and consumers. That's a big reason why Goldman, Sachs & Co. thinks India will be able to sustain 7.5% annual growth after 2005.

Skeptics fear U.S. companies are going too far, too fast in linking up with this giant. But having watched the success of the likes of GE Capital International Services, many execs feel they have no choice. Inside GECIS' Bangalore center—one of four in India—Gauri Puri, a 28-year-old dentist, is studying an insurance claim for a root-canal operation to see if it's covered in a certain U.S. patient's dental plan. Two floors above, members of a 550-strong analytics team are immersed in spreadsheets filled with a boggling array of data as they devise statistical models to help GE sales staff understand the needs, strengths, and weaknesses of customers and rivals. Other staff prepare data for GE annual reports, write enterprise resource-planning software, and process $35 billion worth of global invoices. Says GE Capital India President Pramod Bhasin: "We are mission-critical to GE." The 700 business processes done in India save the company $340 million a year, he says.

Indian finance whizzes are a godsend to Wall Street, too, where brokerages are under pressure to produce more independent research. Many are turning to outfits such as OfficeTiger in the southern city of Madras. The company employs 1,200 people who write research reports and do financial analysis for eight Wall Street firms. Morgan Stanley, J.P. Morgan, Goldman Sachs, and other big investment banks are hiring their own armies of analysts and back-office staff. Many are piling into Mindspace, a sparkling new 140-acre city-within-a-city abutting Bombay's urban squalor. Some 3 million square feet are already leased to Western finance firms. By yearend, Morgan Stanley will fill several floors of a new building.

For Silicon Valley startups, Indian engineers let them stretch R&D budgets. PortalPlayer Inc., a Santa Clara (Calif.) maker of multimedia chips and embedded software for portable devices such as music players, has hired 100 engineers in India and the

U.S. who update each other daily at 9 A.M. and 10 P.M. J.A. Chowdary, CEO of PortalPlayer's Hyderabad subsidiary Pinexe, says the company has shaved up to six months off the development cycle—and cut R&D costs by 40%. Impressed, venture capitalists have pumped $82 million into PortalPlayer.

More Bang for the Buck

Old economy companies are benefiting, too. Engine maker Cummins plans to use its new R&D center in Pune to develop the sophisticated computer models needed to design upgrades and prototypes electronically. Says International Vice-President Steven M. Chapman: "We'll be able to introduce five or six new engines a year instead of two" on the same $250 million R&D budget—without a single U.S. layoff.

The nagging fear in the U.S., though, is that such assurances will ring hollow over time. In other industries, the shift of low-cost production work to East Asia was followed by engineering. Now, South Korea and Taiwan are global leaders in notebook PCs, wireless phones, memory chips, and digital displays. As companies rely more on IT engineers in India and elsewhere, the argument goes, the U.S. could cede control of other core technologies. "If we continue to offshore high-skilled professional jobs, the U.S. risks surrendering its leading role in innovation," warns John W. Steadman, incoming U.S. president of Institute of Electrical & Electronics Engineers Inc. That could also happen if many foreigners—who account for 60% of U.S. science grads and who have been key to U.S. tech success—no longer go to America to launch their best ideas.

Information-technology services could soon account for 7% of India's GDP.

Throughout U.S. history, workers have been pushed off farms, textile mills, and steel plants. In the end, the workforce has managed to move up to better-paying, higher-quality jobs. That could well happen again. There will still be a crying need for U.S. engineers, for example. But what's called for are engineers who can work closely with customers, manage research teams, and creatively improve business processes. Displaced

BRAINPOWER
India and Silicon Valley: Now the R&D Flows Both Ways

The ravages of the dot-com bust are still evident at Andale Inc.'s Mountain View (Calif.) headquarters. Half the office space sits abandoned, one corner of it heaped with discarded cubicle dividers and file cabinets. But looks are deceptive. The four-year-old startup, which offers software and research tools for online auction buyers and sellers, has seen its workforce nearly quadruple in the past year—with most of those jobs in Bangalore.

Andale's 155 workers in India, where employing a top software programmer runs a small fraction of the cost in the U.S., have been the key to the company's survival, says Chief Executive Munjal Shah, who grew up in Silicon Valley. In fact, Indian talent is adding vitality throughout Silicon Valley, where it's getting hard to find an info-tech startup that doesn't have some research and development in such places as Bangalore, Bombay, or Hyderabad. Says Shah: "The next trillion dollars of wealth will come from companies that straddle the U.S. and India."

The chief architects of this rising business model are the 30,000-odd Indian IT professionals who live and work in the Valley. Indian engineers have become fixtures in the labs of America's top chip and software companies. Indian émigrés have also excelled as managers, entrepreneurs, and venture capitalists. As of 2000, Indians were among the founders or top execs of at least 972 companies, says AnnaLee Saxenian, who studies immigrant business networks at the University of California at Berkeley.

Until recently, that brainpower mostly went in one direction, benefiting the Valley more than India. Now, this ambitious disaspora is generating a flurry of chip, software, and e-commerce startups in both nations, mobilizing billions in venture capital. The economics are so compelling that some venture capitalists demand Indian R&D be included in business plans from Day One. Says Robin Vasan, a partner at Mayfield in Menlo Park: "This is the way they need to do business."

The phenomenon is due in no small part to the professional and social networks Indians have set up in the Valley, such as The Indus Entrepreneurs (TiE), in Santa Clara: It now has 42 chapters in nine countries. Prominent Indians such as TiE founder and serial entrepreneur Kanwal Rekhi, venture capitalist Vinod Khosla, and former Intel corp. executive Vin Dham serve as startup mentors and angel investors. In early November, Bombay-born Ash Lilani, senior vice-president at Silicon Valley Bank, led 20 Valley VCs on their first trip to India to scout opportunities. Of the bank's 5,000 Valley clients, 10% have some development work in India, but that's expected to rise to 25% in two years.

Such opportunities for the Valley's Indians flow both ways. Hundreds have returned to India since 2000 to start businesses or help expand R&D labs for the likes of Oracle, Cisco Systems, and Intel.

The downturn—and Washington's decision to issue fewer temporary work visas—accelerated the trend. At a Nov. 6 tech job fair in Santa Clara, hundreds of engineers lined up, résumés in hand, for Indian openings offered by companies from Microsoft Corp. to Juniper Networks Inc. "The real development and design jobs are in India," says Indian-born job-seeker Jay Venkat, 24, a University of Alabama electrical engineering grad.

The deeper, more symbiotic relationship developing between the Valley and India goes far beyond the "body shopping" of the 1990s, when U.S. companies mainly wanted low-wage software-code writers. Now the brain drain from India is turning into what Saxenian calls "brain circulation," nourishing the tech scenes in both nations.

Some Valley companies even credit India with saving them from oblivion. Web-hosting software outfit Ensim Corp. in Sunnyvale relied on its 100-engineer team in Bangalore to keep designing lower-cost new products right through the downturn. "This company would not survive a day if not for the operation in India," says CEO Kanwal Rekhi. Before long, India may prove as crucial to the Valley's success as silicon itself.

—By Robert D. Hof in Santa Clara, Calif., with Manjeet Kripalani in Bombay

technicians who lack such skills will need retraining; those entering school will need broader educations.

Adapting to the India effect will be traumatic, but there's no sign Corporate America is turning back. Yet the India challenge also presents an enormous opportunity for the U.S. If America can handle the transition right, the end result could be a brain gain that accelerates productivity and innovation. India and the U.S., nations that barely interacted 15 years ago, could turn out to be the ideal economic partners for the new century.

With Steve Hamm in New York.

Hints of Comeback for Nation's First Superhighway

CHRISTOPHER MAAG

Most people do not believe that Tim Dufel can push 2,000 tons of steel all the way across New York State. Isn't the old Erie Canal dried up, they ask him, its locks broken, its ditch filled in and forgotten?

They ask these questions even on days like this one, when Mr. Dufel is standing in an orange life vest, watching brown water flood Lock 16 here and lift his loaded barge like a toy battleship in a bathtub.

"Sixty percent of the people I meet have no idea the Erie Canal is even still functioning," Mr. Dufel said. He is assistant engineer on the tugboat Margot and an owner of the New York State Marine Highway Transportation Company, one of the largest shippers on the canal.

After decades of decline, commercial shipping has returned to the Erie Canal, though it is a far cry from the canal's heyday. The number of shipments rose to 42 so far this year during the season the canal is open, from 15 during last year's season, which lasts from May 1 to Nov. 15.

Once nearly forgotten, the relic of history has shown signs of life as higher fuel prices have made barges an attractive alternative to trucks.

"We anticipated we might have an increase in commercial traffic, but nowhere near what we're seeing today," said Carmella R. Mantello, director of the New York State Canal Corporation, a subsidiary of the New York State Thruway Authority that operates the Erie and three other canals.

Along the Erie Canal, business owners who never gave the sleepy waterway much thought are exploring new ways of putting it to use.

"There aren't too many wagon trails left, but we still have the canal," said John Callaghan, a mate on the Margot. "Sure it's history, but it's still relevant. We're making money here."

Completed in 1825, rerouted in parts and rebuilt twice since then, the Erie Canal flows 338 miles across New York State, between Waterford in the east and Tonawanda in the west. It carved out a trail for immigrants who settled the Midwest, and it cemented the position of New York City, which connects with the canal via the Hudson River, as the nation's richest port. In 1855, at the canal's height as a thoroughfare for goods and people, 33,241 shipments passed through the lock at Frankfort, 54 miles east of Syracuse, according to Craig Williams, history curator at the New York State Museum in Albany.

Though diminished in the late 1800s by competition from railroads, commercial shipping along the canal grew until the early 1950s, when interstate highways and the new St. Lawrence Seaway lured away most of the cargo and relegated the canal to a scenic backwater piloted by pleasure boats.

The canal still remains the most fuel-efficient way to ship goods between the East Coast and the upper Midwest. One gallon of diesel pulls one ton of cargo 59 miles by truck, 202 miles by train and 514 miles by canal barge, Ms. Mantello said. A single barge can carry 3,000 tons, enough to replace 100 trucks.

As the price of diesel climbed over $4 a gallon this summer—the national average is now about $3.31 a gallon—more shippers rediscovered the Erie Canal. On one trip in mid-October, the Margot motored down the canal at about seven knots, pushing a barge loaded with a giant green crane. The machine was being transported from Huger, S.C., to the Pinney Dock, operated by the Kinder Morgan Company in Ashtabula, Ohio.

"It really just came down to economics," said Lee Demers, the dock's manager. The other option was to move the crane through the St. Lawrence Seaway, adding more than 1,000 miles and greater fuel costs to the trip.

A few miles east of Little Falls, the canal sat flat as glass, reflecting the orange and red leaves on either bank. As the barge plunged through, the surface curled into a mirrored wave two feet high before breaking into gray chop.

"I've worked the East Coast, the West Coast and the Panama Canal, but up here is some of the most beautiful country you can ever see," said John Schwind, 62, the captain of the Margot, who first learned to pilot tugs here in the 1970s.

A little further on, the barge pitched to port as it rammed muddy sediment along the bottom. After rains flood the canal's tributaries, the mud occasionally becomes so deep that the tug's propellers cannot turn the load, and the barge drifts dangerously close to the bank.

"You have to pay constant attention," Mr. Schwind said. "You can run into trouble real fast."

The canal was dug at least 12 feet deep. But decades of diminished shipping revenue left the canal corporation struggling to keep up with maintenance. Last year, the corporation paid $3 million for a new dredger—its old machine dated from the 1950s.

The agency does not have money for advertising, so this year's growth happened almost entirely by word of mouth. Much of the interest comes from new energy businesses. An old building in Fulton that has been converted into the Northeast Biofuels plant sits on the shore of the Oswego River, which serves as the Oswego Canal and connects to the Erie. The site will also include a carbon dioxide recovery plant, which required moving four large metal tanks there, said Eric Will II, one of the owners of Northeast Biofuels.

Moving the tanks by truck or rail would have required cutting them into pieces and reassembling them at the site, adding tens of thousands of dollars to the cost, Mr. Will said. Instead, the manufacturer delivered the tanks whole by barge.

"It's a nifty thing to do," Mr. Will said, "and it can be very cost-effective."

Some proponents worry that as oil prices drop, the canal could lose its price advantage, and shipping might again slide into oblivion. Others expected this summer's spike to continue because it reminded people that the canal exists.

When a company called Auburn Biodiesel decided to convert an old factory in Montezuma into a biodiesel plant, the building's location beside the canal "was merely an incidental consideration," said David J. Colegrove, the company's president. But after watching the number of cargo shipments along the canal grow, Mr. Colegrove said he hoped to bring soybeans in by barge and use the canal to ship finished product to New York City.

"The amount of money you can save is really eye-popping," he said. "I'm fascinated by the history of the canal, and I'm intrigued by how well it still works."

The Short End of the Longer Life

KEVIN SACK

Throughout the 20th century, it was an American birthright that each generation would live longer than the last. Year after year, almost without exception, the anticipated life span of the average American rose inexorably, to 78 years in 2005 from 61 years in 1933, when comprehensive data first became available.

But new research shows that those reassuring nationwide gains mask a darker and more complex reality. A pair of reports out this month affirm that the rising tide of American health is not lifting all boats, and that there are widening gaps in life expectancy based on the interwoven variables of income, race, sex, education and geography.

The new research adds weight to the political construct popularized by former Senator John Edwards of North Carolina, that there are two Americas (if not more), measured not only by wealth but also by health, and that the poles are growing farther apart.

The most startling evidence came last week in a government-sponsored study by Harvard researchers who found that life expectancy actually declined in a substantial number of counties from 1983 to 1999, particularly for women. Most of the counties with declines are in the Deep South, along the Mississippi River, and in Appalachia, as well as in the southern Plains and Texas.

Ever rising longevity is not a given for all Americans anymore, especially women.

The study, published in the journal PLoS Medicine, concluded that the progress made in reducing deaths from cardiovascular disease, thanks to new drugs, procedures and prevention, began to level off in those years. Those gains, as they shrank, were outpaced by rising mortality from lung cancer, chronic obstructive pulmonary disease and diabetes. Smoking, which peaked for women later than for men, is thought to be a major contributor, along with obesity and hypertension.

"Some people are actually sinking," said Majid Ezzati, one of the report's authors. "The line of excuse that we can live with inequality as long as no one is getting worse is just no longer there."

The researchers found statistically significant declines for women in 180 of the 3,141 counties in the United States and in 11 counties for men. In an additional 783 counties for women

and 48 for men, there were declines that did not reach the threshold of statistical significance.

Of particular concern is that the gap in life expectancy between top and bottom counties expanded by two years for men and by about 10 months for women. In the worst-performing counties, all in southwestern Virginia, the drop in life expectancy over the 16-year period was nearly six years for women and two and a half years for men. In the counties showing the greatest improvement, many in the desert West, life expectancy rose nearly five years for women, and nearly seven years for men.

The first of the two reports, released two weeks ago by the Congressional Budget Office, declared that the life expectancy gap is growing between rich and poor and between those with the highest and lowest educational attainment, even as it is narrowing between men and women and between blacks and whites.

Pointing to the effects of smoking, obesity and chronic disease, the budget analysts wrote that "in recent decades, socioeconomic status has become an even more important indicator of life expectancy, whether measured at birth or at age 65." Among the implications, they wrote, is that Social Security payroll taxes will become less progressive as the wealthy increase their longevity advantage over the poor.

Peter R. Orszag, the budget office's director, said that the decline in life expectancy among some Americans was "remarkable in an advanced industrial nation" and that he believed the growing gap related to income inequality. "We've had sluggish income growth at the bottom and rapid income growth at the top for the last three decades," he said.

Mr. Edwards said in an interview that the new findings on disparities demonstrate both the reach and consequences of income inequality. "The wealth and income disparity effectively infiltrates all parts of people's lives," he said.

What remains to be determined by the increasingly dynamic field of research into health disparities is precisely how income interacts with factors like race, gender and education to give some people better odds of living longer.

A growing life span was an American birthright for generations, but no more.

Taken to their extreme, the numbers can be striking: a 2006 study found that Native American men in southwestern South Dakota could expect to live to 58, while Asian women in Bergen County in New Jersey had a life expectancy of 91.

For some groups at some times, disparities can widen and shrink because of societal changes (like fluctuating homicide rates) and medical developments (like the emergence of H.I.V., or the discovery of drugs to treat it). But the causes of more lasting trends may not always be obvious, and some research suggests that income alone cannot explain away many differences.

For example, a 2006 study found that low-income whites in the northern Plains could expect to live four years longer than low-income whites in Appalachia and the Mississippi Valley. Other research indicates that health insurance status, which can relate directly to income, may not be a significant determinant of longevity.

And yet, this month's Harvard study showed that counties with declining or stagnant life expectancy were poorer than those with improving numbers. Recent cancer studies have found that the uninsured are more likely to fail to get a diagnosis until late stages of the disease. Research also shows that many of the behaviors that drive mortality—unhealthy diet, smoking, poor management of chronic disease—are more common among low-income Americans.

"We know from hundreds of studies that income does have an impact on health, but it's not a simple relationship," said Sam B. Harper, an epidemiologist at McGill University who has studied the issue.

Dr. Ezzati, of the Harvard School of Public Health, asked: "How much of this is pure material well being, the ability to purchase high-quality food, the ability to have a particular lifestyle? And how much of it is the impact of income on risk behaviors like alcohol and tobacco and stress mechanisms that are more psychosocial? There's a series of debates around that that are unresolved."

As for a prescription, Dr. Ezzati and his colleagues are realists. In a 2006 study, they concluded that "because policies aimed at reducing fundamental socioeconomic inequalities are currently practically absent in the U.S.," life expectancy disparities would have to be addressed through public health strategies directed at reducing the risk factors that cause chronic disease and injuries.

Dr. Ezzati noted that few industrialized countries have had declines of comparable duration. "This is a very unusual pattern," he said, "and the question we're starting to ask is, 'Is the fact that the bottom 20 percent is not getting better, and may be worse off, going to drive the health of the whole country?'"

Never Too Late to Scramble: China in Africa

The emperor's new clothes? China is rapidly buying up Africa's oil, metals and farm produce. That fuels China's surging economic growth, but how good is it for Africa?

In his office in Lusaka, Zambia's capital, Xu Jianxue sits between a portrait of Mao Zedong and a Chinese calendar. His civil-engineering and construction business has been doing well and, with the help of his four brothers, he has also invested in a coal mine. He is bullish about doing business in Zambia: "It is a virgin territory," he says, with few products made locally and little competition. He is now thinking of expanding into Angola and Congo next door. When he came in 1991, only 300 Chinese lived in Zambia. Now he guesses there are 3,000.

Mr Xu reflects just a tiny part of China's new interest in Africa. This year alone many bigger names than his have come visiting. Li Zhaoxing, China's foreign minister, swept through west Africa in January; President Hu Jintao visited Nigeria, Morocco and Kenya in April; and the prime minister, Wen Jiabao, knocked off seven countries in June. In the first week of November Chinese and more than 30 African leaders will gather at the first Sino-African summit in Beijing. And Chinese companies, most of them owned by the state, have been marching in the footsteps of their political leaders. But is this all good for Africa? Is it bringing the trade and investment that Africa so badly needs, or just meddling and exploitation?

The summit in Beijing is being greeted by Chinese officials and the country's state-run media with an effusion reminiscent of the cold-war era, when China cosied up to African countries as a way of demonstrating solidarity against (Western) colonialism and of outdoing its ideological rival, the Soviet Union. It supported African liberation movements in the 1950s and 1960s, and later built railways for the newly independent countries, educated their students and sent them doctors.

China's main aim then was to gain influence. Now China wants commodities more than influence. Its economy has grown by an average of 9% a year over the past ten years, and foreign trade has increased fivefold. It needs stuff of all sorts—minerals, farm products, timber and oil, oil, oil. China alone was responsible for 40% of the global increase in oil demand between 2000 and 2004.

The resulting commodity prices have been good for most of Africa. Higher prices combined with higher production have helped local economies. Sub-Saharan Africa's real GDP increased by an average of 4.4% in 2001–04, compared with 2.6% in the previous three years. Africa's economy grew by 5.5% in 2005 and is expected to do even better this year and next.

Which countries are the main beneficiaries? For copper and cobalt, China looks to the Democratic Republic of Congo and Zambia; for iron ore and platinum, South Africa. Gabon, Cameroon and Congo-Brazzaville supply it with timber. Several countries in west and central Africa send cotton to its textile factories.

Oil, however, is the biggest business. Nigeria, Africa's biggest oil-producer, has been getting lots of attention. CNOOC, a state-owned Chinese company, paid $2.7 billion in January to obtain a minority interest in a Nigerian oilfield, and China recently secured exploration rights in another four. In Angola, which has now overtaken Saudi Arabia as China's biggest single provider of oil, another Chinese company is a partner in several blocks. China has shown similar interest in other producers such as Sudan, Equatorial Guinea, Gabon and Congo-Brazzaville, which already sells a third of its output to Chinese refiners.

Just the Beginning

As a result, trade between China and Africa has soared from $3 billion in 1995 to over $32 billion last year. But China's commerce with the world also expanded over the same period, so Africa makes up only 2.3% of the total. This constitutes about 10% of Africa's total trade.

However, trade between China and Africa is expected to double by 2010. Although Europe remains Africa's main partner, its share has melted from 44% to 32% of the region's foreign trade within the past ten years, whereas America's share has, like China's, risen. For some countries, the redirection of exports has been dramatic. China now takes over 70% of Sudan's exports, compared with 10% or so in 1995. Burkina Faso sends a third of its exports, almost all of which are cotton, to China, compared with virtually nothing in the mid-1990s. China is now Angola's largest export market after the United States.

Africa has found more than a new buyer for its commodities. It has also found a new source of aid and investment. According to China's statistics, it invested $900m in Africa in 2004, out of the $15 billion the continent received. This was a huge increase, though most of it went to oil-producing countries. But its aid is spread more widely. It has cancelled several billion dollars of African debt, which has helped to build roads, railways, stadiums and houses in many countries.

This largesse is sometimes an entry ticket. In Nigeria China's promises to invest about $4 billion in refineries, power plants and agriculture were a condition for getting oil rights. In Angola a $4 billion line of low-interest credit enables Chinese companies to help rebuild the bridges, roads and so on that were destroyed in decades of war. The debt is repaid in oil.

Fewer Complications

For Angola, which has been keen to get going with the reconstruction of its infrastructure, China's straightforward approach is an attractive alternative to the persnicketiness of the IMF and the Paris Club of creditors, which have been quibbling over terms for years. So it is with many African countries, fed up with the intrusiveness of Europeans and Americans fussing about corruption or torture and clamouring for accountability. Moreover, the World Bank and many Western donors were until recently shunning bricks-and-mortar aid in favour of health and education. China's credit to Angola is not only welcome in itself. It has reduced the pressure from the West.

Thanks to China, therefore, workers from the Middle Kingdom in straw hats are now helping Angolans to lay down new rails on the old line from Luanda to the eastern province of Malange. Another railway, from Benguela to Zambia, once used to carry copper, is also being rebuilt. China is happy: the work helps offset some of its trade deficit with Angola. The Angolan government is also happy: it is rebuilding its shattered economy at last.

For Jose Cerqueira, an Angolan economist, China is welcome because it eschews what he sees as the IMF's ideological and condescending attitude. "For them," he says, "we should have ears, but no mouth." Others are pleased because China is ready to pass on some of its technology. It is, for example, helping Nigeria to launch a second satellite into space. Some African officials, disillusioned with the Western development model, say that China gives them hope that poor countries can find their own path to development.

And Now the Snags

The love affair with China, however, may be sour as well as sweet. For countries that do not sit on oil or mineral deposits, higher commodity prices make life harder. Even for producers there are risks. A recent report by the World Bank argues that Africa's new trade with China and India opens the way for it to become a processor of commodities and a competitive supplier of cheap goods and services to Chinese and Indian consumers. But another report, from the OECD, a club of industrialised countries, argues that China's appetite for commodities may stifle producers' efforts to diversify their economies. Oil rigs and mines create few jobs, it points out, and tend to suck in resources from other industries. And if Africa is to escape its vulnerability to the capricious movements of world commodity prices, it must start to export more manufactures. On this the World Bank adds its own warning: China and India must end their escalating tariffs on Africa's main exports.

China is also bringing irresistible—some say unfair—competition to Africa. All over Africa Chinese traders can now be seen selling cheap products from the homeland, not just electronics but plastic goods and clothes. In Kamwala market in Lusaka a host of Chinese shops have appeared over the past couple of years. "Two years ago," says Muhammad, a local trader of Indian origin, "I did not have time to sit down; now I'm sitting doing nothing." Though his shelves are full of clocks and radios made in China, he blames his enforced idleness on the competition brought by Chinese traders.

Zambian and other African consumers do not share his despondency. They like Chinese prices. But in some countries consumers are less well organised than textile workers, and in South Africa the trade unions have succeeded in getting the government to negotiate quotas on Chinese clothing imports. Still, the power of China's productivity and economies of scale—never mind government subsidies—certainly hurts local industries. Textile factories in places like South Africa, Mauritius and Nigeria have been badly hit. In tiny Lesotho, where making clothes for Europe or America is the only industry around, this has been catastrophic.

The working conditions, as well as the prices, set by Chinese employers are also a concern to some Africans. The alleged ill-treatment of workers in a Chinese-owned mine in Zambia in July led to a violent protest in which several workers were shot. And many Chinese firms bring in much of their own labour, rather than hiring locals. China brought in thousands of its own workers to build the 1,860km (1,160-mile) Tazara railway between Lusaka and the Tanzanian port of Dar es Salaam in the 1970s. It was finished ahead of schedule, but Tanzania and Zambia still have to rely on Chinese technical help to maintain it. African hopes of technology transfer may be over-optimistic.

Human Rights Are Optional

Some say China's involvement will erode efforts to promote openness and reduce corruption, especially in oil and mining. Nigeria insists that Chinese companies must respect its new anti-graft measures, and the latest bidding round for oil blocks in Angola has been the most open so far. In both countries it is unclear whether China's presence is making corruption better or worse. It is clear, though, that China is not interested in pressing African governments to hold elections or be more democratic in other ways. That helps to explain why China directs so much money towards Sudan, whose odious regime can count on China's support when resisting any UN military intervention in Darfur. China invested almost $150m in Sudan in 2004, three times as much as in any other single country. When American and Canadian oil companies packed their bags there, China quickly stepped in, drilling wells and building pipelines and roads. The Chinese are supposed to be building an armaments factory as well.

China's lack of interest in human rights is something that President Robert Mugabe of Zimbabwe can also be thankful for. Shunned by the West, and with his country's economy in a shambles, he has turned to China for political and economic support—and got it. After he launched Operation Murambatsvina last year, in which 700,000 people had their homes or businesses destroyed, China neutered all attempts at discussion, let alone condemnation, in the UN Security Council. However, despite this, China may not want to squander any more money in a country that has no oil and few mineral rights left to dispose of.

But China's friendship and support at the UN comes with one important political string attached: the endorsement of the one-China policy. To date 48 African countries have paid due obeisance to Beijing: Chad, Senegal and Liberia are the latest to have abandoned their recognition of Taiwan. The suggestion by Michael Sata, the main opposition candidate in Zambia's presidential election on September 28th, that he would have recognised Taiwan if he had won was enough to bring the first public intervention by China in the internal affairs of an African country; the ambassador said that China would consider cutting diplomatic relations if Mr Sata won (which he did not).

That is a warning to Africans that this new interloper in their continent is no more altruistic than its predecessors. Still, that does not mean China's involvement is bad and it is certainly not to be stopped. It is up to Africans to ensure that they get a fair deal from it. If so, both China and its African partners can be winners.

Where Business Meets Geopolitics

A vast new pipeline that will bring Caspian oil westwards has opened, at a cost of $4 billion. Western countries see it as important in reducing their reliance on oil from Russia and the Gulf states. Like other pipelines in the region its route marks the shifting of political allegiances that has driven a wedge between Moscow and Washington.

When Tsar Nicholas I laid plans for a railway between Moscow and St Petersburg 150 years ago, he took the direct route. By laying a ruler on the map he decreed a straight connection between Russia's two great cities, bar a small kink where it is said that he accidentally drew around his finger—timid courtiers failed to alert him to his detour.

Today's most contentious oil and gas pipelines, mainly sited or planned in and around Russia, are not susceptible to such autocratic whim. On Wednesday May 25th, the 1,800 km Baku-Tbilisi-Ceyhan (BTC) pipeline officially opened for business some 13 years after its conception and at a cost of $4 billion. The pipeline, built by a consortium led by Britain's BP, will bring Caspian oil from Azerbaijan across the Caucasus to the Mediterranean coast of Turkey, from where it will be tankered to markets worldwide. When it is fully operational it could transport 1m barrels a day, just over 1% of the world's current oil consumption.

The strategic value of Caspian oil did not escape Adolf Hitler. He fatally over-extended his army's supply lines with a dash to secure the region's oil reserves, resulting in a decisive defeat at Stalingrad. Russia kept control of the region's oil until the break-up of the Soviet Union. Then western governments and oil companies, searching for fresh sources of oil in a bid to reduce reliance on the Middle East, advanced on the Caspian themselves. But the region has failed to live up to its early promise. Azerbaijan is never likely to become the new Kuwait. America's Energy Information Administration estimates that the Caspian region has oil reserves of between 17 billion and 33 billion barrels, rather than the 200 billion touted in the mid-1990s, plus a fair bit of natural gas.

Despite this disappointment, more of this oil and gas is coming on tap and new means of getting it to market are required. Unlike many other big oil producers, such as Saudi Arabia and it neighbours clustered around the Persian Gulf, Caspian oil and gas is landlocked and set apart from the sea by a selection of countries with differing ambitions and loyalties. Tsar Nicholas and his planning techniques have thus given way to complex geopolitical [maneuvering].

The quickest route to the sea is to go south through Iran. But handing control of a key pipeline to such an unpredictable regime was inconceivable. Instead, in 2001, a 1,510 km pipeline (known as CPC) opened between the Tengiz oilfields of Kazakhstan and the Russian port of Novorossisk. This pipe, built by a consortium led by ChevronTexaco, was the first privately owned pipeline crossing Russian soil (though its government has a 24% stake). Russia, keen to retain control over oil exports, which it uses as a foreign-policy lever, has kept the rest of its 48,000 km of oil pipelines strictly under the control of state-run Transneft.

The BTC pipeline, though the most expensive option for exporting Caspian oil, was backed by America because it avoided Russia, thereby reducing the dependence of the Caucasus and Central Asia on Russian pipelines. The pipeline also provided an opportunity to bolster regional economies that the West is courting, especially those of Georgia, Azerbaijan and Turkey, a NATO ally, and build support for America in the region. Georgia's location gives it a "strategic importance far beyond its size," according to America's State Department.

Upgrading an alternative route through Georgia to Supsa on the Black Sea would have made for a far shorter (and cheaper) pipeline. But Turkey complained that it would lead to an unsustainable level of shipping passing through the Bosporus Strait that bisects Istanbul. At Washington's urging, the BTC pipeline wended its complex way through Azerbaijan, Georgia and Turkey. However, some critics of the pipeline point out that the oil revenues provided to Azerbaijan will help to prop up the country's autocratic and corrupt regime. And environmentalists have complained that the pipe slices through a national park in Georgia.

Pipe Hype

Oil-thirsty China is also keen to get its hands on the region's resources, and in September work began on a 1,000 km pipeline from Atasu in central Kazakhstan to western China. Eventually,

China hopes to extend the pipe another 2,000 km across Kazakhstan to the Caspian oilfields. However, plans to link it with existing Russian pipelines to allow the movement of Siberian oil through Kazakhstan are likely to be stymied by Transneft.

In Europe, too, countries are not afraid to indulge in a little pipeline politics. In February, Ukraine gave Russia a slap in the face by agreeing to reverse the flow of the Odessa-Brody pipeline. This was supposed to take Russian oil south to the Ukrainian Black Sea port of Odessa and then on to world markets by tanker; it will now pump Caspian oil up through Ukraine and into the European pipeline network. Ukraine has recently experienced fuel shortages, which its prime minister blamed on over-reliance on Russian oil.

A little further afield, two other proposed pipelines demonstrate how the quest for energy resources can overcome some political difficulties while creating others. A thawing of relations between India and Pakistan prefigured a recent announcement that India is considering a 2,775 km gas pipeline link to Iran. America fiercely opposes involvement with Iran and wants to use outside access to Iranian energy resources as a lever to stop the country's nuclear programme. Russia has broadly supported Iran's nuclear programme, in part as a counter to America's ambitions in the Caspian region.

For its part, Pakistan is party to an agreement to build a gas pipeline from Turkmenistan, which has substantial gas fields, through Afghanistan to the Pakistani coast. The Taliban, still an active threat in Afghanistan, has vowed to disrupt the project if American firms are involved. But the threat of terrorism, even in these unstable parts of the world, is probably exaggerated. Although insurgents have disrupted exports of Iraqi oil by blowing up pipelines there, most pipes run underground and attacks on them have not been common. Other oil installations, such as refineries and depots, are generally thought to be at greater risk.

In fact, the vast majority of oil pipelines are uncontentious and unthreatened: there is 322,000 km of plumbing carrying crude oil, gas and refined products in America alone, for instance. But as the race for oil and gas from remote areas hots up, threatening old alliances and forging new ones, the politics of pipeline placement will become ever more complicated. The days of the tsar and his ruler are over.

Article 21

Half-Way from Rags to Riches

**Vietnam has made a remarkable recovery from war and penury, says
Peter Collins. But can it change enough to join the rich world?**

Knees and knuckles scraping the ground, the visitors struggle to keep up with the tour guide who is briskly leading the way through the labyrinth of claustrophobic burrows dug into the hard earth. The legendary Cu Chi tunnels, from which the Viet Cong launched waves of surprise attacks on the Americans during the Vietnam war, are now a popular tourist attraction. Visitors from all over the world arrive daily at the site near the city that used to be called Saigon, renamed Ho Chi Minh City after the Communists took the south in 1975.

Alongside the wreckage of an abandoned M41 tank another friendly guide demonstrates a dozen types of improvised booby-traps with sharp spikes that were set in and around the tunnels to maim pursuing American soldiers. The Vietnamese not only welcome the tourist dollars Cu Chi brings in, but are also rather proud of it. They feel it demonstrates their ingenuity, adaptability, perseverance and, above all, their determination to resist much stronger foreign invaders, as the country has done many times down the centuries.

These days Vietnam also has plenty of other things to be proud of. In the 1980s Ho Chi Minh's successors as party leaders damaged the war-ravaged economy even more by attempting to introduce real communism, collectivising land ownership and repressing private business. This caused the country to slide to the brink of famine. The collapse soon afterwards of its cold-war sponsor, the Soviet Union, added to the country's deep isolation and cut off the flow of roubles that had kept its economy going. Neighbouring countries were inundated with desperate Vietnamese "boat people."

Since then the country has been transformed by almost two decades of rapid but equitable growth, in which Vietnam has flung open its doors to the outside world and liberalised its economy. Over the past decade annual growth has averaged 7.5%. Young, prosperous and confident Vietnamese throng downtown Ho Chi Minh City's smart Dong Khoi street with its designer shops. The quality of life is high for a country that until recently was so poor, and its larger cities have retained some of their colonial charm, though choking traffic and constant construction work are beginning to take their toll.

An agricultural miracle has turned a country of 85m once barely able to feed itself into one of the world's main providers of farm produce. Vietnam has also become a big exporter of clothes, shoes and furniture, soon to be joined by microchips when Intel opens its $1 billion factory near the capital, Hanoi. Imports of machinery are soaring. Exports plus imports equal 160% of GDP, making the economy one of the world's most open.

All this has kept government revenues buoyant despite cuts in import tariffs. The recent introduction of company taxes is also helping to fill the government's coffers. Spending on public services has surged, yet public debt, at an acceptable 43% of GDP, has remained fairly stable.

Having made peace with its former foes, Vietnam hosted Presidents Bush, Putin and Hu at the Asia-Pacific summit in 2006 and joined the World Trade Organisation in 2007. This year it has one of the rotating seats on the UN Security Council.

Vietnam's Communists conceded economic defeat 22 years ago, in the depths of a crisis, and brought in market-based reforms called *doi moi* (renewal), similar to those Deng Xiaoping had introduced in China a few years earlier. As in China, it took time for the effects to show up, but over the past few years economic liberalisation has been fostering rapid, poverty-reducing growth.

The World Bank's representative in Vietnam, Ajay Chhibber, calls Vietnam a "poster child" of the benefits of market-oriented reforms. Not only does it comply with the catechism of the "Washington Consensus"—free enterprise, free trade, sensible state finances and so on—but it also ticks all the boxes for the Millennium Development Goals, the UN's anti-poverty blueprint. The proportion of households with electricity has doubled since the early 1990s, to 94%. Almost all children now attend primary school and benefit from at least basic literacy.

Vietnam no longer really needs the multilateral organisations' aid. Multilateral and bilateral donors together have promised the country $5.4 billion in loans and grants this year, but with so much foreign investment pouring in, Vietnam's currency reserves increased by almost double that figure last year. At least the aid donors have learned from the mid-1990s, when excessive praise discouraged Vietnam from continuing to reform, prompting an exodus of investors. Now the tone in private meetings with officials is much franker, says a diplomat who attends them.

Vietnam has become the darling of foreign investors and multinationals. Firms that draw up a "China-plus-one" strategy for

The Making of Modern Vietnam

Date	Event
1959–65	Gradual build-up to Vietnam war
1968	Communists launch Tet Offensive against American and South Vietnamese forces
	My Lai massacre by American troops
1969	President Richard Nixon announces gradual troop withdrawal
1973	Pans peace accords signed
1975	Communists capture south and reunite country
1979	Vietnam invades Cambodia and topples Pol Pot. Brief war with China ensues
1986	*Doi moi* economic liberalisation launched
	Pro-Moscow diplomatic policy replaced with "friends everywhere" doctrine
1989	Number of "Vietnamese boat people" fleeing penury hits peak
1991	The Soviet Union, Vietnam's cold-war sponsor, disintegrates
1994	America lifts trade embargo
	UN brokers pact on resettling and repatriating Vietnamese boat people
1995	Vietnam joins ASEAN
	America and Vietnam restore diplomatic relations
2000	Vietnam creates stockmarket and legally recognises private enterprises
	America-Vietnam trade pact signed
2006	Vietnam hosts Asia-Pacific summit
2007	Vietnam joins World Trade Organisation
2008	Vietnam wins temporary seat on UN Security Council

Source: *The Economist*

new factories in case things go awry in China itself often make Vietnam the plus-one. Wage costs remain well below those in southern China and productivity is growing faster, albeit from a lower base. When the UN Conference on Trade and Development asked multinationals where they planned to invest this year and next, Vietnam, at number six, was the only South-East Asian country in the top ten.

The government's programme of selling stakes in publicly owned firms and exposing them to market discipline has recently gathered pace. At the same time the switch from a command economy to free competition has allowed the Vietnamese people's entrepreneurialism to flourish. Almost every household now seems to be running a micro-business on the side, and a slew of ambitious larger firms is coming to the stockmarket.

Much of the praise now being showered anew on the country is deserved. The government is well on course for its target of turning Vietnam into a middle-income country by 2010. Its longer-term aim, of becoming a modern industrial nation by 2020, does not seem unrealistic.

But from now on the going may get tougher. As Mr Chhibber notes, few countries escape the "middle-income trap" as they become richer. They tend to lose their reformist zeal and see their growth fizzle. A study in 2006 by the Vietnamese Academy of Social Sciences concluded that further reductions in poverty will require higher growth rates than in the past because the remaining poor are well below the poverty line, whereas many of those who recently crossed it did not have far to go.

The Stench of Corruption

The Communist Party leadership openly admits that the Vietnamese public is fed up with the endemic corruption at all levels of public life, from lowly traffic policemen and clerks to the most senior people in ministries. In 2006, just before the party's five-yearly congress, the transport minister resigned and several officials were arrested over a scandal in which millions of dollars of foreign aid were gambled on the outcome of football matches. The leadership insists it is doing its best to clean up, but a lot remains to be done.

Almost as bad as the corruption is the glacial speed of legislative and bureaucratic processes. Proposed laws have to pass through all sorts of hoops before taking effect, with endless rounds of consultations to build consensus. The dividing line between the Communist Party, the government and the courts is not always clear. The justice system is rudimentary. Lawyers have no formal access to past case files, so they find it hard to use precedent in legal argument.

The government is part-way through a huge project to slim the bureaucracy and streamline official procedures. It recently cut the number of ministries from 28 to 22. Yet for the moment the bureaucratic logjam is stopping the country building the roads, power stations and other public works it needs to maintain its growth rate. Nguyen Tan Dung, the prime minister, says that if growth is to continue at its current rate, the country's electricity-generating capacity needs to double by 2010. That seems a tall order, to put it mildly.

Soaring car-ownership is leaving the country's underdeveloped roads increasingly gridlocked. In an admirably liberal attempt to limit price distortions as oil surged above $100 a barrel, the government slashed fuel subsidies in February. But one effect will be to stoke inflation, already worryingly high at 19.4% in March. Bank lending surged by 38% last year as firms and individuals borrowed to speculate on shares and property.

The government is finding it much harder to manage an economy made up of myriad private companies, banks and investors than to issue instructions to a limited number of state institutions, especially as the public sector is currently suffering a drain of talent to private firms that are able to offer much higher pay.

What Could Go Wrong

All this leaves Vietnam's continued economic development exposed to a number of risks:

- Rising inflation—which is hurting low earners in particular—and a growing shortage of affordable housing could create a new urban underclass among unskilled workers who have left the land for the cities.

Combined with rising resentment at official corruption and the increasing visibility of Vietnam's new rich, this could cause social friction and bring strikes and protests, chipping away at the political stability that has underpinned Vietnam's strong growth and investment.

- Trade liberalisation and increased domestic competition will benefit some firms and farmers but hurt others—especially inefficient state enterprises. These could join forces and press the government to halt or even reverse the reforms.

- The slumping stockmarket or perhaps a property crash could cause a big firm or bank to fail. Given the country's weak and untested bankruptcy laws and financial regulators, the authorities may find it hard to deal with that kind of calamity.

- Natural disasters, from bird flu to floods, could cause chaos.

- The economy could come up against the limits of its creaking infrastructure and the shortage of people with higher skills. Jammed roads, power blackouts and the inability to fill managerial and professional jobs could all bring Vietnam's growth rate crashing down.

Vietnam has set itself such demanding standards that even if some combination of these factors did no more than push annual growth below 5%, it would be seen as a serious setback. The foreign minister, Pham Gia Khiem, notes that Vietnam's current growth of around 8-9% is lower than that in Asia's richest economies at the same stage in their development.

Despite the risks ahead, Vietnam has already provided the world with an admirable model for overcoming war, division, penury and isolation and growing strongly but equitably to reach middle-income status. This model could be followed by many impoverished African states or, closer to home, perhaps by North Korea. If it can be combined with gradual political liberalisation, it might even offer something for China to think about.

Malaria

It's Not Neglected Any More (But It's Not Gone, Either)

Hellen Gelband

Public consciousness about malaria is rising in this country. Just a few years ago, many Americans thought that malaria was an ancient plague and were surprised to discover that it still plagues populations in Africa, Asia, and other tropical parts of the globe. Of course, people still have much to learn, but between the President and First Lady visiting malaria control programs in Africa and the Denver Nuggets raising money for bed nets treated with mosquito-zapping insecticide, malaria is harder to miss these days.

Here are the rote statistics: half a billion cases and one million dead each year—most of them African children. What's new—and startling, certainly for the global health community—is the fact that a billion dollars is now pouring into malaria control every year. Less than a decade ago, it was just a few tens of millions. With such an enormous financial commitment and the attention of the world, will this investment pay off? Will we finally be able to change the numbers on the malaria scoreboard? People are talking big: the "e word" is in play again. *Eradication.* Is it a pipe dream or can it be reality?

While scientific advances in the treatment of malaria are cause for optimism, the lack of a unified worldwide plan or vision for malaria control remains a serious concern. More people sleep under protective nets and have access to effective drugs than ever before but malaria-endemic countries tend to be among the world's poorest, which also means they have the weakest healthcare infrastructures. And while malaria may be the most important health problem historically, it is overshadowed by AIDS—which not only makes people vulnerable to other diseases, but also has soaked up the best and the brightest in the healthcare workforce.

Can we control malaria? Or will it continue to control the lives of the people affected? A lot depends on what happens over the next few years: if success can be documented, funding will probably continue to flow. But, if progress is not great enough, despite the large sums devoted to tackling malaria, the disease may win again.

Today's Control Measures Can Work

Clear evidence has emerged, from the places where the current wave of malaria control started earliest, that the tools we have do work. The big three interventions are effective drugs, insecticide-impregnated bednets, and the spraying of indoor walls with insecticide (referred to as indoor residual spraying). Take Kwazulu Natal (KZN), the state that had the highest malaria burden in South Africa up to the year 2000, when the KZN malaria control program conducted house-to-house campaigns of indoor spraying, and switched to the best type of drug. Prior to that, some districts reported 5,000 cases each month during the high season. In 2001, the numbers fell to 1,000 per month. Since 2002, not more than a few hundred cases have been reported. Today, mothers no longer spend their days caring for children in crowded malaria wards. Both the annual number of cases and number of deaths in KZN have fallen 90 percent. Zanzibar, a large island off the Tanzanian coast, and other countries (Rwanda and Ethiopia, for instance) where insecticide-treated nets are integral to the mix, are beginning to yield similar stories.

We still need better drugs and insecticides, and the search continues for the holy grail: a vaccine against malaria. But we know that using current methods will lead to huge declines in the malaria burden. Whether or not they can lead to eradication is still an open question.

Funding Is at an All-Time High

The harsh reality is that the best science and the best intentions will have little impact without funding. That goes for implementing malaria control programs and for carrying out research needed to advance knowledge, both in the laboratory and in the field. Recent progress has been possible because of money. Current funding for malaria control is at an all-time high and still in crescendo mode. Since 2000, three major new funding sources have transformed the scene: the Global Fund for AIDS, Tuberculosis and Malaria (the Global Fund), the World Bank Booster Program for Malaria, and the President's Malaria Initiative (through the U.S. Agency for International Development, USAID). The Department for International Development, the British bilateral aid agency, is also a major donor to malaria efforts and, in smaller amounts, other countries have increased aid as well. The role of the Bill & Melinda Gates Foundation, in money and in visibility for malaria, cannot be overlooked. Overall, it adds up to a billion dollars per year.

The Global Fund has made the biggest financial contribution over the largest number of countries, and has the best chance of maintaining a long-term commitment. The President's initiative is billed as a five-year, $1.2 billion program, and like the President's Emergency Plan for AIDS, funding will likely be renewed if progress is being made. It is difficult to project what priorities may look like in the United States five years from now, however. Clearly, a different president will be in office who may want his or her stamp on some other cause.

Better Tools on the Way

The first and only serious attempt to eradicate malaria globally, which began with much fanfare in 1955, succeeded in southern Europe and large parts of Asia and the Americas, but failed in sub-Saharan Africa. The World Health Organization's (WHO) malaria eradication campaign relied on a single tool—spraying of the then-remarkable insecticide DDT. By 1969, when a halt was called to the campaign, it was clear that DDT alone could not wipe out malaria in Africa, where intensity of transmission was higher (year-round in many areas) and infrastructure was poor. Most obviously, DDT-resistant

mosquitoes took over well before the job was done. Where DDT had outlasted the species that spread malaria elsewhere, the African vector (*Anopheles gambiae*) was tougher, and in the end, mosquitoes triumphed. Some also believe that sub-Saharan Africa was written off as a lost cause for malaria, and that sufficient effort was not made.

> **The first and only serious attempt to eradicate malaria globally, in the mid-20th century, succeeded in southern Europe and large parts of Asia and South America, but failed in sub-Saharan Africa. . . . It was clear that DDT alone could not wipe out malaria in Africa, where intensity of transmission was higher and infrastructure was poor. Some also believe that sub-Saharan Africa was written off as a lost cause for malaria, and that sufficient effort was not made.**

It could be different this time. We have a bigger and better arsenal of tools and, equally important, a better understanding of how they work. We know from well planned and executed field trials that insecticide-treated bed nets reduce childhood deaths from malaria. Net technology itself has improved: an earlier generation required users to retreat them every three months with insecticide, but the current models incorporate insecticide into the fabric itself. And we have a new generation of drugs—artemisinin-combination therapies, or ACTs—that are even more effective than chloroquine, which was lost to resistant malaria parasites after a decades-long run. Even DDT has been rehabilitated. The years during which it was not used has winnowed out the resistant mosquitoes and DDT is now used more judiciously, by spraying only internal walls, as in KZN. A few other insecticides can also be used, but development of new insecticides has lagged.

For the long term, the malaria drug pipeline is fuller than it's ever been. Although novel drugs may come from a variety of sources, the Medicines for Malaria Venture, a non-profit "public-private partnership," has the deepest and broadest inventory of drugs in development of any organization. Over time, even the best new drugs will need replacement—not in crisis, but as a matter of course. That should now be possible, although it will likely be another decade before the partnership's R&D results in new forms of treatment.

Malaria Knowledge Is Advancing

The breadth and organization of knowledge are also advancing in important ways. Recently, the first results of the Malaria Atlas Project (MAP) were published, combining sophisticated data processing and old-fashioned, shoe-leather epidemiologic detective work. The international Kenya-based MAP team (including David Smith, an RFF visiting scholar) has produced the most detailed malaria map to date. Using records unearthed from around the globe, it shows not just how many people are at risk of malaria, but also their level of risk. MAP could be the basis of a global plan for malaria control, containment, and eventual eradication. Talk is now about "shrinking the map."

Drug Resistance: Liability and Opportunity

One of the biggest threats to malaria control is drug resistance. The world was lucky that chloroquine—the 20th century mainstay—was effective for decades. For reasons not well understood, very few malaria parasites ever maintained genetic mutations conferring true resistance to this drug. But over time, the progeny of a resistant strain from Southeast Asia finally spread throughout Asia and then Africa. In Asia, replacement drugs were used starting in the 1960s. By the 1990s, mortality rates in Africa were rising because chloroquine no longer worked, and African countries, by and large, did not have the resources to switch drugs. The exception was a switch to another remarkably inexpensive drug, sulfadoxine-pyrimethamine (SP). It was very effective initially but, unlike chloroquine, was rendered ineffective in a few short years by drug-resistant malaria.

Chloroquine and SP resistance were both global catastrophes and wake-up calls: malaria drugs are precious, shared resources that must be managed so that they do the most human good, but they also must be protected from loss to drug resistance for as long as possible. The fact that the world is now relying on one drug class—the artemisinins—as the backbone of malaria drug treatment for at least the next decade makes protection all the more imperative.

Continuing research at RFF is playing a key role in advancing both science and policy for better stewardship of antimalarial drugs. This spring, RFF researchers hosted scientists and policymakers from around the world at a first-of-a-kind conference on antimalarial treatment strategies, held in South Africa. A major theme was that malaria drugs are shared resources, and their effectiveness, a "global public good."

The conference was the culmination of 18 months of work that extended earlier epidemiologic modeling at RFF. The earlier work predicted large benefits from using malaria drugs in combination (rather than as monotherapy, which had been the norm), both in terms of saving lives and prolonging the effectiveness of drugs. The current combinations all include an artemisinin plus a companion drug (ACTs)—each of which should be effective malaria drugs for the locale.

Would using more than one combination in a given population give even greater protection to the drugs? Would they remain effective for years, maybe even decades, longer? That is just what the models developed at RFF predict: multiple first-line therapy should significantly delay the spread of resistant parasites. But can endemic countries implement such policies?

No one expects a clinic or doctor to randomly assign patients to one ACT or another when they come in needing treatment. So RFF has suggested practicable alternatives: children get one ACT and adults another, for example. Or the use of one ACT in the public sector and another in the private sector. Today, multiple drugs are sold from big-city pharmacies down to small village shops. Unfortunately, many are ineffective (people still buy chloroquine and SP because they are affordable and, currently, ACTs are not), substandard, or outright counterfeits. Both the affordability and the quality of drugs sold in the public sector are another focus of RFF work.

Financing Malaria Drugs through "Radio Malaria"

"Radio Malaria" is the nickname for the Affordable Medicines Facility-malaria—AMFm. The AMFm strategy was born of the central idea in a 2004 study by an Institute of Medicine (IOM, part of the U.S. National Academy of Sciences) committee. RFF Senior Fellow Ramanan Laxminarayan was a member of the committee, and I served as the study director. The committee's idea for a global subsidy has developed into the outlines of an international organization, slated to begin operation in 2009. By the plan's outline, manufacturers of high-quality ACTs (judged, as currently, by the WHO or another international authority) will sell them at "chloroquine" prices to governments and to the wholesalers that supply the private markets in endemic countries. AMFm will then pay the manufacturer a supplemental amount for each dose sold, so that manufacturers will be paid a fair (but competitive) wholesale price.

Money, effective control measures, knowledge, innovative financing mechanisms, the promise of even better interventions—all are on the increase where malaria is concerned. . . . The key to future worldwide eradication will be a plan with global scope that can shrink the malaria map until it no longer exists.

By taking advantage of a chain of distribution that already exists through the private sector (where more than half of antimalarials are acquired currently), AMFm would expand access to these lifesaving drugs and delay their loss of effectiveness to resistant malaria parasites for years or decades. The title of the IOM report says it all: *Saving Lives, Buying Time.*

Money, effective control measures, knowledge, innovative financing mechanisms, the promise of even better interventions—all are on the increase where malaria is concerned. Thus far, interventions and plans have been approached in nearly all cases on a country-by-country basis. Some countries have seen greater success than others. The key to future worldwide control—possibly even eradication—will be a plan with global scope that can shrink the malaria map until it no longer exists.

Tsunamis: How Safe Is the United States?

THOMAS AARON GREEN

Introduction

The tsunami that occurred in the Indian Ocean on September 26, 2004 raised global awareness of the vulnerability of coastal populations to tsunami hazards. Natural hazards such as earthquakes, tsunamis, and hurricanes occur globally without any discrimination. There was no warning system in place for countries surrounding the Indian Ocean when the undersea earthquake occurred, and questions are being raised as to how safe the United States coastline is even though it does have a warning system. Many geographical factors contribute to a complex set of answers to this particular question. This paper examines the intricate relationships of geographic factors such as location of fault lines, coastal population density, varying velocities of tsunami travel, height of waves, urban infrastructures, evacuation plans, time available for evacuation, and the degree of devastation.

The Asian Tsunami

The absence of tsunami advance warning systems in the Indian and Atlantic Oceans proved to be extremely disastrous when one of the worst tsunamis ever recorded hit. The lack of warning and inadequate knowledge of what warning signs to look for caused people of Southeast Asia to remain unaware of the events that would follow the undersea earthquake registered at 9.0 on the Richter scale. In light of the tsunami that struck Southeast Asia, a report by Horton (2005) states that it has caused leaders around the world to support an advanced warning system in the Indian Ocean in hopes of preventing another disaster of this magnitude. The Australian Broadcasting Corporation (ABC) reported a death toll estimated at 286,000 on January 31, 2005, with a projected count of over 300,000. The Toronto Star produced a business report by Stuart Laidlaw (2005) estimating the economic loss from the tsunamis at around $13 billion. However, Laidlaw stressed that insurance companies have yet to accurately and completely calculate the actual cost of the widespread damage, as Canadian corporations are sending aid while trying to make contact with Canadian workers who reside in the affected areas. Making matters worse, the areas that were hit the hardest by the series of tsunamis are primarily third world countries, which possess inadequate

medical facilities in comparison to first world countries. The lack of clean water has caused more problems, as reported by the Anjana Pasricha (2005). Epidemics of malaria and other water-carried diseases became major health concerns, while another related worry is the psychological trauma faced by the people of the region.

The Likelihood of a Tsunami Striking the U.S.

The occurrence of tsunamis striking the United States has been reported by Roberts in his article on *The History of a Tsunami* (Roberts, 1961), in which he discusses the damage from a tsunami that happened from a rock slide on July 9, 1958. Moreover, the geographic location that is in danger of a tsunami as destructive as the one that happened in Southeast Asia is likely to be off the coast of Washington, Oregon, and California. The Geological Survey of Canada and the Sidney Subdivision of Emporia State University both stress that the Cascadia Subduction Zone is geomorphically identical to the subduction zone off the west coast of Sumatra, Indonesia. The most hazardous section of the Cascadia Subduction Zone is the Juan de Fuca fault line, a tectonic plate that is moving under the North American plate. The Juan de Fuca fault line is responsible for the volcanic activity that occurs throughout the Cascade Mountain region, which runs from British Columbia, Canada, to Northern California. The map in Figure 1 shows that the main area of subduction occurring between the North American plate and Pacific plate is along the Washington and Oregon coastline.

The impact of the Pacific plate subducting under the North American plate will cause a tsunami. Brain Atwater, a U.S. Geological Surveyor discovered that a major earthquake did occur off the coast of Washington State along the Juan de Fuca fault line in 1700 triggering a tsunami (Nash, 2005). It caused serious physical damage to the then uninhabited Washington coastline.

Today, the Washington coastline is heavily populated, especially in the Puget Sound region, which is the location of great metropolitan areas such as Seattle-Tacoma, Washington and Vancouver-Victoria, British Columbia. In fact, there are many fault lines in or in close proximity to the Puget Sound, and the chance of

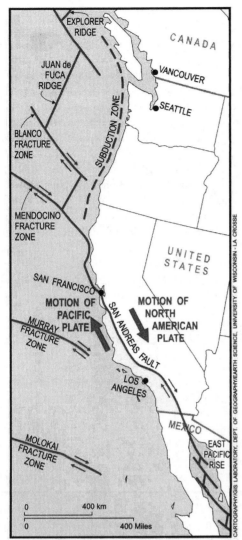

CARTOGRAPHY/GIS LABORATORY, DEPT. OF GEOGRAPHY/EARTH SCIENCE, UNIVERSITY OF WISCONSIN-LA CROSSE

Figure 1 Areas of tsunami hazards along the U.S. West Coast.

a major earthquake occurring here and causing a tsunami is a real possibility. Gonzalez (1999) reports that about 300,000 people live or work in nearby hazardous coastal regions, and at least as many tourists travel through these areas every year. He further states that a major earthquake off the coast of Washington on the Juan de Fuca fault or in the Puget Sound, as suggested by the U.S. Geological Survey, would leave little to no time for people to evacuate, even with an operating advanced tsunami warning system.

How Did the Asian Tsunami Occur?

The U.S. Geological Survey reported that the earthquake which occurred off the coast of Indonesia was initially reported to be an 8.1 on the Richter scale; later seismology tests and data revised it to 9.0. The earthquake was triggered by a slip in the Indo-Australian and Eurasian plates' subduction zone about ten miles

off the coast of Indonesia on a sub-plate known as the Burma plate. The Tsunami Laboratory in the Geophysics Department at the University of Washington stated that prior to any alarm of danger, these two plates were previously wedged eighteen miles under the sea floor and ten miles off the coast of Indonesia. The Indo-Australian plate was moving up against the Burma plate, causing pressure to build as neither one could move. Eventually, the pressure was too great, and these two plates snapped. The snapping of these two plates caused seismic waves to move outward from the epicenter, creating a magnitude 9.0 earthquake. The force of the physical movement of the snapping plates set ocean waters in motion as tsunami waves.

Some Historical Background of Tsunamis

Roberts (1961), in an annual report for the Smithsonian Institute, provided a synopsis of the Alaskan landscape and how it was scraped clean of vegetation from massive tsunamis. In some cases, evidence from photographs which are featured in the report is more than 40 years old, indicating that tsunamis, depending on their magnitude, will leave evidence of damage for decades or longer. Both the Geophysics Department at the University of Southern California (2004) and Roberts (1961) reported that more than twenty tsunami events have impacted the State of California in the past two centuries. Tsunamis are not rare phenomena in the Pacific Ocean according to the U.S. Geological Survey, but they are less frequent in the Atlantic and Indian Oceans. This is because the fault lines along the Atlantic and Indian Ocean floors are spreading further apart while the fault lines in the Pacific are converging together. Converging fault line zones are much more susceptible to tsunamis.

The Basic Physics of a Tsunami

The word *tsunami* is a Japanese word that means harbor wave (or sometimes called a seismic wave) because of the destruction they cause to cities, towns, and rural landscapes along coastal regions. They typically hit land in a series of three or more waves. The speed and height of these waves depend on the magnitude of release of sea floor seismic waves from the epicenter (National Tsunami Hazard Mitigation Program, 2002). The Columbia Encyclopedia (2001) states that the time frame between the first and second seismic waves will vary. According to the National Oceanic and Atmospheric Administration (NOAA), the second seismic wave could strike five minutes to one hour after the first. The speed at which tsunamis travel depends on the depth of the ocean. A large time gap can increase the death toll due to the wait between the first and second tsunamis. Many people continue to go about their business, not realizing that a second and third wave are to follow. The tsunami that struck Hilo, Hawaii on April 1, 1946 was triggered by an earthquake off the Alaskan coast. While the people in Hilo continued with their daily lives, they were not expecting a second wave which struck five hours after the first, killing 159 people and costing $26 million in damages. The Hilo tsunami disaster prompted the installation of a warning system and as a result, the Pacific Tsunami Warning Center was founded in 1948 (Kong, 2003; NOAA, 2004).

Is the United States More Prepared and Safer in the Event of a Tsunami?

The question whether the United States is more prepared for a tsunami than Southeast Asia is difficult to answer. The first tsunami warning system established in the U.S. in 1948 was embryonic. With further scientific work, the U.S. National Tsunami Hazard Mitigation Program developed and deployed the Deep-Ocean Assessment and Reporting of Tsunamis system (D.A.R.T.) in 1995 and further refined it in 1998, Figure 2. The purposes of D.A.R.T. are twofold: to reduce the loss of life and property in U.S. coastal communities and eliminate false alarms and the high economic cost of unnecessary evacuations (Milburn, 1996). The D.A.R.T. system has already been tested enough in the past years to prove that it can successfully monitor any seismic movement on the ocean floor.

D.A.R.T. is not a perfect system for the entire U.S.; it currently provides coverage only for the Pacific coasts of the U.S. (Milburn, 1996), The way that the D.A.R.T. system works is through the use of a bottom pressure recorder (BPR), which has the ability

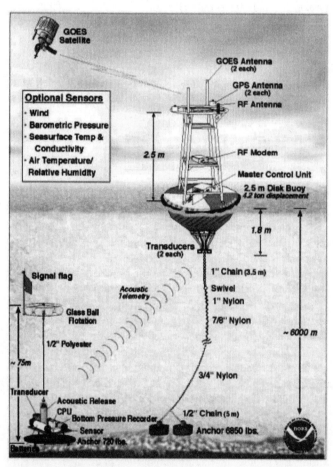

Figure 2 The Deep-Ocean Assessment and Reporting of Tsunami mooring system (D.A.R.T.).

Source: National Oceanic and Atmospheric Administration (NOAA).

to detect a tsunami as small as one centimeter. The tsunami wave readings are then transferred to a buoy on the seas surface via an acoustic link. The buoy transmits the data to a satellite and subsequently to ground stations. Computers in ground stations then decipher the mathematical equations and send the text data to the National Oceanic and Atmospheric Administration's Tsunami Warning Centers. In the case of the Indonesia Tsunami of December 2004, the first waves hit the Indonesian coast of Aceh in about 10–15 minutes after the earthquake. NOAA reported that they had a speed of 34 miles per hour and were 50 feet high. Even if D.A.R.T. had been placed in Indonesia, there would not have been enough time to evacuate the heavily populated Indonesian coast of Aceh Province. The major causes of the high mortality rate from this tsunami were due to heavy population along the coastlines and the short distance of Aceh Province from the epicenter. The absence of a warning system and an evacuation plan did not help matters. These are important geographic reasons why residents living along earthquake-prone coastlines are in constant threat of danger.

Mitigating Possible Tsunamis in the U.S.

The National Tsunami Hazard Mitigation Program has begun developing mitigating programs to help educate people living in communities with tsunami hazards. They should know the degree of vulnerability their geographic region has, plan newer development further inland on higher grounds to avoid future losses, and most importantly, plan an evacuation.

The D.A.R.T. system is a well-designed, advanced set of instruments, but geographic factors determine the magnitude of destruction, as well as loss of life. D.A.R.T. provides a warning as soon as an undersea earthquake occurs, however, the location or proximity to the epicenter determine whether a coastal community will have sufficient time to evacuate. This means an advanced warning system can save lives, particularly for populations further away from the epicenter. Tsunamis are still unpredictable and can occur on any day and any time of the day or night. An advanced tsunami warning system is obviously necessary. But even though, such a system may be work to record and provide warning as the event occurs, there is no guarantee that it will offer a sufficient time frame to successfully complete an evacuation. To help put realistic geographic factors into perspective, consider the following three scenarios.

Scenario 1: The Washington and Oregon Coast

The state of Washington is at the heart of the Cascadia Subduction Zone, which is responsible for all of the volcanic activity that has occurred and will continue to occur in the future. The 9.0 earthquake that occurred in this region in 1700 caused minimal cultural destruction and perhaps very little loss of life as the population then was much less than that of today. The damaging effects on the physical landscape left by the tsunami are still visible to those with seismology or geomorphology training. About 50 miles off the coast of Washington is where the Juan

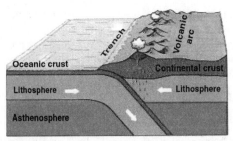

Figure 3 The subduction process, showing the convergence of the oceanic and continental plates.

Source: NOAA.

de Fuca plate is slowly subducting under the North American Plate. Subduction, a process where one tectonic plate is sliding under another, generally creates the largest earthquakes in terms of magnitude and causes the most devastation (U.S. Geological Survey, 2003b), see Figure 3.

In this Washington/Oregon coast earthquake scenario, the earthquake will be plotted along the Juan de Fuca fault line, eighteen miles under the ocean floor. The likelihood of a magnitude 9.0 earthquake is possible as the Juan de Fuca fault line is part of the Cascadia Subduction Zone, meaning that the Pacific plate is sliding under the North American plate (Figure 1). Eventually, the two tectonic plates will become wedged, unable to move, as rock does not move and grind well against rock. Pressure will start to build as a result of the wedging that has occurred near these two plate boundaries. The pressure will eventually build to a maximum, resulting in a break, sending seismic shockwaves to the surface. The effects of the pressure buildup and ultimate breakage will cause water on the ocean floor to be pushed out from the epicenter, creating a tsunami (U.S. Geological Survey, 2003c; Freeman, 1999). The tsunami will move from the epicenter in all directions at an average speed of 500 miles per hour. Since the Washington coastline is the closest to the pressure zone, an evacuation opportunity will only be a period of 10–15 minutes. As the tsunami approaches land, the shallowing depth of the beach profile will cause the rate of motion of the tsunami to slow down, drastically increasing the height of the wave possibly to over 50 feet (Milburn, 1996; Yurkovich, & Howell, 2003). The strength of the earthquake is also a determining factor on a tsunamis rate of motion. The U.S. Geological Survey, Kong (2003), and Roberts (1961) have all concluded that tsunamis come in a series of three waves. In the event of a 9.0 earthquake at the Juan de Fuca Fault, a 50-foot wave will likely hit the Washington coast. This will be the first of three; the second and third waves are always the worst. The first wave will hit at a reduced speed of about 34 miles per hour and the force of this wave will cause flooding as far as two miles inland. If the epicenter directly faces the Columbia River's inlet, the resulting tsunami would indirectly create more dangerous effects. The magnitude of the earthquake would cause the Columbia River to rise several feet, forcing ocean waves into the river and causing heavy flooding as far inland as Portland, Oregon.

The level of flooding along the Pacific coast and Columbia River banks will cause problems for the area and for the United States as a whole. First, the Columbia River is a major shipping route for the Pacific Northwest. People of this region rely on the Columbia River to transport goods out of Portland to other places,

especially Seattle, San Francisco, and Anchorage. This does not even include their international trade with Japan or elsewhere in the Pacific Rim. The economic problems that occur would not only be felt by the people of the Pacific Northwest, but would impact the rest of the United States. The Columbia River is the heart of the Pacific Northwest, as the river is used for hydroelectricity, and the Columbia and Snake Rivers carry 17 million tons of cargo annually to and from the Pacific Ocean (Foundation of Water and Energy Education, 2000). The Center for Coastal and Land-Margin Research (1996) already stated that the occurrence of a tsunami hitting the Washington coastline and causing harm to the Columbia River is a real possibility. NOAA and the U.S. Geological Survey both assert that the damage which would occur from an earthquake and the following tsunamis will cause havoc to the coastline, wash away small coastal communities, and damage ship loading docks in addition to the loss of lives. Along the coastal areas of Washington and Oregon alone, the number of deaths can reach 40,000–50,000 simply because there would not be enough time to evacuate the population. A 9.0 magnitude earthquake could level roadways and reduce the number of escape routes, limiting the effectiveness of any evacuation. The destruction of bridges and roads and the loss of soil and vegetation cover will make it harder for the residents along the coastal areas to recover. In some cases, most of these small communities will be completely destroyed. The natural ecological conditions will require decades to recover.

Scenario II: Puget Sound

Figure 4 shows the density of seismic fault lines within Puget Sound. The City of Seattle is situated on the banks of Puget Sound, with four fault lines crossing through its metro area. This map shows the tsunami risks for Seattle/Tacoma in the event of an earthquake due to the proximity to the fault lines. If an earthquake were to take place inside Puget Sound instead of occurring in the Pacific off the Washington coastline, the results would be catastrophic because the evacuation time would be much less than the 10–15 minutes tsunami travel time in Scenario I where there is at least 50 miles between the Juan de Fuca Fault and the Washington coast. The Puget Sound area is known as the *Crustal Zone* or the *Third Source Zone* (U.S. Geological Survey, 2003b, May 7). Both University of Washington and U.S.G.S. scientists concluded that Puget Sound is shrinking about half an inch a year in a north to south direction. They do not have an explanation for this shrinkage but believe that the Seattle fault is responsible for the shortening of the Puget Sound. Shrinkage generally indicates pressure-building along fault lines.

The main concern today is the destruction of major metropolitan areas, home to 3.5 million inhabitants in Greater Seattle alone (The World Almanac, 2004), more if the tsunami reaches Canadian cities such as Vancouver and Victoria. A 9.0 magnitude earthquake would not only level a considerable part of Seattle, but could bring a tsunami at an estimated speed of 500 miles per hour and wave heights of 135 feet within minutes and cause flooding two miles inland (Roberts, 1961). In addition, the shape of Puget Sound is such that it will act as a funnel in many directions, trapping water in a confined area, and not being able to dissipate water into the open seas. In the case of the Hilo tsunami, the fact that the City of Hilo is located at the end of a funnel-shaped

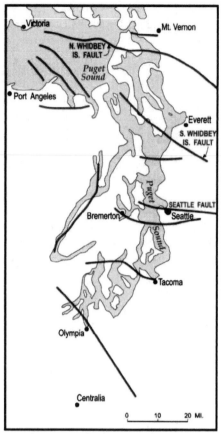

CARTOGRAPHY/GIS LABORATORY, DEPT. OF GEOGRAPHY/EARTH
SCIENCE, UNIVERSITY OF WISCONSIN-LA CROSSE

Figure 4 Numerous fault lines are located directly below Puget Sound.

bay directly facing the epicenter of the Alaskan quake caused the waves to rise even higher than usual. Tsunamis generated from inside Puget Sound could also create similar funnel-effects causing waves to rise even higher.

Greater Seattle is the major metropolitan area of the Pacific Northwest, home to some of the most widely known corporations in the U.S., such as Boeing, Microsoft, and the retail outlet giant, Nordstrom. The tsunami effects brought on by an earthquake within Puget Sound would bring an estimated cost of $20 billion dollars in damages. A tsunami from a 9.0 magnitude earthquake would cripple the city, causing economic loss to the geographically linked economic hinterlands that are dependant on the Seattle economy. Seattle accounts for nearly 87% of Washington State's total exports and is a major bub for transporting goods domestically and internationally. Almost all Seattle commerce would be directly or indirectly affected if loading docks and ports were destroyed. Furthermore, Seattle has more residents living along its shoreline than San Diego and Miami Beach; the death toll in the event of a tsunami could be in the 10,000–15,000 range (Seattle Chamber of Commerce, 2005).

Scenario III: San Francisco, California

San Francisco is located on a peninsula that protrudes out from the state of California. The city and surrounding areas have a dense population and reside over a complex series of fault lines.

The main fault line is the famous San Andreas Fault, one of the most active faults in the world. The San Andreas Fault extends through the entire state of California, geologically separating a piece of the state on the Pacific plate and the remainder on the North America plate, causing the entire state to be geologically unstable. In addition, the area of California on the Pacific plate is slowly drifting northward while the rest of California on the North American plate is slowly drifting southward, San Francisco is geographically located on top of the San Andreas Fault and several of its connected tributaries. For this reason, the city is one of the most unstable places on earth.

San Francisco is a major economic giant and key financial center for the western United States and the world. It is home to 40 international Fortune 500 companies as well as thousands of medium and smaller businesses, which constitute San Francisco's thriving economy. Its population has a rich diversity, and they are especially proud of having no leading ethnicity. Globally, it is located between London and Tokyo in the international market and between San Diego and Seattle in the Pacific market. This centrality allows it to be a major financial center, with thousands of companies, making San Francisco the New York City of the West Coast (Sims, 2000). Given its economic importance regionally, nationally, and internationally, it is important to recognize that a major earthquake would cause massive economic loss. San Francisco processes billions of dollars in transactions through its banking centers and markets on a daily basis. A tsunami generated from a 9.0 magnitude earthquake occurring off the coast over the Santa Luca Banks Fault could bring costs of damage estimated upwards of 20 billion dollars.

Because of the fifteen- to twenty-mile proximity of the Santa Luca Banks Fault to San Francisco, a tsunami could travel up to one mile inland, with flooding occurring at least a half mile further. Since the epicenter will also be in close proximity to San Francisco and the California coastline in general, there is a likely probability that the first tsunami will hit the Bay Area as few as five minutes after the earthquake. The speed of the tsunami will likely be around 500 miles per hour, with waves reaching 135 feet high, hitting everything in its wake one to two miles inland.

The death toll in the Bay City will be high. Even with the D.A.R.T. system a 5–15 minute warning is not enough time to evacuate a heavily populated city. Once again, the earthquake will likely destroy numerous transportation routes and bridges, bringing traffic to a standstill. Not only will San Francisco sustain serious loss of life from the earthquake, but also from second and third wave tsunamis that will follow after in undetermined time frames.

A city of this size and wealth (with a population of over 7 million) built on a major fault line will continue to be at risk for a major seismic disaster. In the event of a 9.0 earthquake from the Santa Luca Banks Fault, the city could be leveled from the earthquake alone. A tsunami is unavoidable if the earthquake is caused by subduction.

Conclusion

Cities and coastal areas in close proximity to subduction processes along fault lines are most prone to disasters. The advanced seismic instrumentation developed and used as a tsunami warning system and understanding the geographies of tsunamis will play a major role in mitigating the damage and lose of life caused by them. The D.A.R.T. warning system is a useful tool and needs to be placed throughout the world. While the sophisticated D.A.R.T. system may not have helped those who perished in Aceh Province in Indonesia, it would have helped residents of India and Sri Lanka since the epicenter was located much further away from them allowing a relatively longer time for them to evacuate.

Damage from tsunamis such as the Asia Tsunami of 2004 are not only limited to loss of lives and destruction of property. They have longer lasting effects on the economy of these places, on the ecological systems along coastal areas, on the psychological well-being of coastal communities, and on governments and planners who face the decisions of having to rebuild or relocate entire coastal populations. These are difficult tasks to address. It takes years and decades to recover from such devastations. The United States has witnessed the social, environmental, and human losses resulting from the Katrina hurricane disaster in the U.S. Gulf Coast. A tsunami disaster occurring along the western coast of the U.S. has the potential for even greater destruction, dependent on the geographic location of the occurring epicenters of earthquakes which generate tsunamis.

References

Australian Broadcasting Corporation (ABC) Asia Pacific. (2005) *Asian tsunami death toll continues to rise*. Retrieved: 2005, February 11. http://web.ask.com/redir?U=http%3a%2f%2ftm.wc.ask.com

Center for Coastal and Land-Margin Research. (1995–1996). *Science for Society: Impact of tsunamis on Oregon coastal communities*. Retrieved: 2005, February 23. www.ccalmr.ogi.edu/tsunami/

City and County of San Francisco. 1999–2005. Retrieved: 2005, January 24. http://www.ci.sf.ca.us/

Columbia Encyclopedia, The (Sixth Edition). 2001, *Tsunami*. Columbia University Press. Retrieved: 2005, February 11. http://www.bartleby.com/65/ts/tsunami.html

Foundation of Water & Energy. (1999–2000). *What Makes the Columbia River Basin unique and how we benefit*. Retrieved: 2005, February 23. www.fwee.org/c-basin.html

Freeman, Linda. 1999. *Cascadia Subduction Zone*. Emporia State University, Retrieved: 2005, January 17. http://www.snowcrest.net/freemanl/siskiyou/cascadia/cascadia.html

Geological Survey of Canada-Sidney Subdivision. (2003, June 28), *Earthquake Processes: Cascadia Subduction Zone*. Retrieved: 2005, January 17. http://www.pgc.nrcan.gc.ca/geodyn/docs/slip/slip.infol.html

Gonzalez, Frank, 1999, May, 18. *Scientific American: Tsunami*. Retrieved: 2005, January 11. http://www.sciam.com/printversion.cfm

Horton, Richard. *The Lancet*. (2005, January 15.) Threats to human survival: A WIRE to warn the world. (Vol. 365). p.191–193.

Kong, Laura. (2003). Oceanography: Special Report. *2004 Science Year, The World Book Annual Science Supplement*.

Laidlaw, Stuart. (2005, January 4). Canadian firms scramble in Asian tsunami zone. Business Reporter. *The Toronto Star*. Retrieved: 2005, February 11. www.thestar.com/NASApp/cs/ContentServer?pagename=thestar

Milburn, H.B. September 1996. Deep-Ocean Assessment and Reporting of Tsunamis. National Oceanic and Atmospheric Administration. http://www.pmel.noaa.gov/tsunami/Dart/dartpbl.html

Nash, J. Madeleine. (2005, January 17), An American Tsunami? National Tsunami Hazard Mitigation Program, *The Designing for Tsunamis: Sewn Principles for Planning & Designing for Tsunami Hazards*. 2001, March.

National Oceanic and Atmospheric Administration. (2004). *NOAA Backgrounder: NOAA and Tsunamis*. Washington D.C. FEMA.

North American at Night. (1995–2005). Photo. Retrieved: 2005, February 12. http://www.art.com/asp/sp-asp//PD--10091850/NorthAmericaatNight.asp

Pasricha, Anjana. (2005, January 13). World Health Organization. *Tsunami Health Risks Remain*. Retrieved: 2005, February 11. http://www.Voanews.Com/english/2005-01-13-voal7.cfm

Roberts, Elliot B. *History of a Tsunami*. 1961. Washington: Smithsonian Publication. p.327–340.

San Francisco *Chamber of Commerce: Economic Growth*. 2005, January 24. http://www-spur.org/documents/sims.pdf

Seattle Chamber of Commerce. (1995–2005). Office of Economic Development. http://www.seattle.gov/EconomicDevelopment/

Sims, Kent. *San Francisco Economy: Implications for Public Policy*. 2000, July 10. San Francisco Planning And Urban Research Association, www.spur.org

Tsunami Laboratory. Institute of Computational Mathematics and Mathematic Geophysics. University of Washington. 2005, January 13. Retrieved: 2005, January 14. http://www.cwis.Usc.edu/

U.S. Geological Survey. U.S. Department of the Interior. (2003, October 15a). *Preliminary Earthquake Report*. Earthquake Hazards Program. Retrieved: 2005, February 12. http://earthquake.usgs.gov/recenteqsww/Ouakes/usslav.htm

U.S. Geological Survey. (2003, May 7b). Pacific Northwest Geologic Mapping and Urban Hazards. Retrieved: 2005, February 24. http://wrgis.wr.Usgs.Gov/docs/wgmt/pacnw/lifeline/eqhazards.html

U.S. Geological Survey. U.S. Department of the Interior. (2002, December 31). *Eastern Region Geography*. Retrieved: 2005, January 17. http://mac.usgs.gov/isb/pubs/booklets/elvadist/elvadist.html

U.S. Geological Survey. U.S. Department of the Interior. (2004, December 24). *Tectonic Summary*. National Earthquake Information Center. Earthquake Hazards Program. Retrieved: 2005, February 12. http://neic.us gs.gov/neis/bulletin/neicslav_ts.html

U.S. Geological Survey. U.S. Department of Interior. (2003, October 20). *Historic Earthquakes & Earthquake Statistics*. Earthquake Hazards Program. Retrieved: 2005, February 12. http://earthquake.Usgs.gov/faq/hist.html

The World Almanac and Book of Facts. (2004). Metropolitan Areas, p. 375. New York City.

Yalciner, A.C., Pelinovsky, E., Okal, E., Synolakis, C.E. (eds.). (2003). Producing Tsunami Inundation Maps: The California Experience. *Submarine Landslides And Tsunamis*, p. 315–326. Netherlands: Kluwer Academic Publishers.

Yurkovich, E.S. & Howell, D.G. (November 2003). Analysis of Hazard, vulnerability, population, and infrastructure. *Abstracts with Programs Geological Society of America*, 35(6), p. 18.

UNIT 4

Spatial Interaction and Mapping

Unit Selections

Key Points to Consider

• Describe the spatial form of the place in which you live. Do you live in a rural area, a town, or a city, and why was that particular location chosen?

• How does your hometown interact with its surrounding region? With other places in the state? With other states? With other places in the world?

• What are the geopolitical consequences of an ice-free Arctic Ocean?

• What problems occur when transportation systems are overloaded?

• In what ways have telecommunication and the Internet brought world regions "closer" to each other?

• How can a balance be achieved between economic development and environmental conservation?

• Why do the regions of the South and West continue to draw the lion's share of U.S. migrants?

• How good a map-reader are you? Why are maps useful in studying a place?

Student Website
www.mhcls.com

Internet References

Capitol Region Council of Governments (CRCOG)
www.crcog.org/gissearch
Edinburgh Geographical Information Systems
http://www.geo.ed.ac.uk/home/gishome.html
Geography for GIS
http://www.ncgia.ucsb.edu/cctp/units/geog_for_GIS/GC_index.html
GIS Frequently Asked Questions and General Information
http://factfinder.census.gov/home/saff/main.html

International Map Trade Association
http://www.maptrade.org
PSC Publications
http://www.psc.isr.umich.edu
Telegraph.co.uk
http://www.telegraph.co.uk/travel/picturegalleries/
Worldmapper
www.worldmapper.org

© Comstock/Getty Images

Geography is the study not only of places in their own right but also of the ways in which places interact. Highways, airline routes, telecommunication systems, and even thoughts connect places. These forms of spatial interaction are an important part of the work of geographers.

Richard Florida's article uses four maps to counter Thomas Friedman's notion that the world is flat. *Fortune Magazine* maps produced during World War II are reviewed in the next selection. "Sea Change . . ." deals with major changes in the Arctic stemming from climate change. Cartograms are used in the next article to illustrate global demographic shifts. "Deaths Outnumber Births . . ." uses maps to show population decline in the Midwest and Appalachia. The next article concludes that the "one person, one vote" notion in the United States is simply not true. "Calling All Nations" uses a thematic map to indicate the connectivity of the United States via phone calls with other countries in the world.

The map of corn production and ethanol processing shows a strong spatial relationship between the two activities. "Clogged Arteries" points to severe congestion on U.S. highways. "Manifest Destinations" is a one-page summary of current U.S. migration patterns, economic concentrations in metro areas, population distributions, and aspects of urban sprawl. The last article discusses the continued devastation of AIDS in southern Africa.

It is essential that geographers be able to describe the detailed spatial patterns of the world. Neither photographs nor words could do the job adequately, because they literally capture too much of the detail of a place. Therefore, maps seem to be the best way to present many of the topics analyzed in geography. Maps and geography go hand in hand. Although maps are used in other disciplines, their association with geography is the most highly developed.

A map is a graphic that presents a generalized and scaled-down view of particular occurrences or themes in an area. If a picture is worth a thousand words, then a map is worth a thousand (or more!) pictures. There is simply no better way to "view" a portion of Earth's surface or an associated pattern than with a map.

The World Is Spiky

Globalization has changed the economic playing field, but hasn't leveled it.

RICHARD FLORIDA

The world, according to the title of the *New York Times* columnist Thomas Friedman's book, is flat. Thanks to advances in technology, the global playing field has been leveled, the prizes are there for the taking, and everyone's a player—no matter where on the surface of the earth he or she may reside. "In a flat world," Friedman writes, "you can innovate without having to emigrate."

Friedman is not alone in this belief: for the better part of the past century economists have been writing about the leveling effects of technology. From the invention of the telephone, the automobile, and the airplane to the rise of the personal computer and the Internet, technological progress has steadily eroded the economic importance of geographic place—or so the argument goes.

But in partnership with colleagues at George Mason University and the geographer Tim Gulden, of the Center for International and Security Studies, at the University of Maryland, I've begun to chart a very different economic topography. By almost any measure the international economic landscape is not at all flat. On the contrary, our world is amazingly "spiky." In terms of both sheer economic horsepower and cutting-edge innovation, surprisingly few regions truly matter in today's global economy. What's more, the tallest peaks—the cities and regions that drive the world economy—are growing ever higher, while the valleys mostly languish.

The most obvious challenge to the flat-world hypothesis is the explosive growth of cities worldwide. More and more people are clustering in urban areas—the world's demographic mountain ranges, so to speak. The share of the world's population living in urban areas, just three percent in 1800, was nearly 30 percent by 1950. Today it stands at about 50 percent; in advanced countries three out of four people live in urban areas. Map A shows the uneven distribution of the world's population. Five megacities currently have more than 20 million inhabitants each. Twenty-four cities have more than 10 million inhabitants, sixty more than 5 million, and 150 more than 2.5 million. Population density is of course a crude indicator of human and economic activity. But it does suggest that at least some of the tectonic forces of economics are concentrating people and resources, and pushing up some places more than others.

Still, differences in population density vastly understate the spikiness of the global economy; the continuing dominance of the world's most productive urban areas is astounding. When it comes to actual economic output, the ten largest U.S. metropolitan areas combined are behind only the United States as a whole and Japan. New York's economy alone is about the size of Russia's or Brazil's, and Chicago's is on a par with Sweden's. Together New York, Los Angeles, Chicago, and Boston have a bigger economy than all of China. If U.S. metropolitan areas were countries, they'd make up forty-seven of the biggest 100 economies in the world.

Unfortunately, no single, comprehensive information source exists for the economic production of all the world's cities. A rough proxy is available, though. Map B shows a variation on the widely circulated view of the world at night, with higher concentrations of light—indicating higher energy use and, presumably, stronger economic production—appearing in greater relief. U.S. regions appear almost Himalayan on this map. From their summits one might look out on a smaller mountain range stretching across Europe, some isolated peaks in Asia, and a few scattered hills throughout the rest of the world.

Population and economic activity are both spiky, but it's innovation—the engine of economic growth—that is most concentrated. The World Intellectual Property Organization recorded about 300,000 patents from resident inventors in more than a hundred nations in 2002 (the most recent year for which statistics are available). Nearly two thirds of them went to American and Japanese inventors. Eighty-five percent went to the residents of just five countries (Japan, the United States, South Korea, Germany, and Russia).

Worldwide patent statistics can be somewhat misleading, since different countries follow different standards for granting patents. But patents granted in the United States—which receives patent applications for nearly all major innovations worldwide, and holds them to the same strict standards—tell a similar story. Nearly 90,000 of the 170,000 patents granted in the United States in 2002 went to Americans. Some 35,000 went to Japanese inventors, and 11,000 to Germans. The next ten most innovative countries—including the usual suspects in Europe plus Taiwan, South Korea, Israel, and Canada—produced roughly 25,000 more. The rest of the broad, flat world accounted for just five percent of all innovations patented in the United States. In 2003 India generated 341 U.S. patents and China 297. The University of California alone generated more than either country. IBM accounted for five times as many as the two combined.

This is not to say that Indians and Chinese are not innovative. On the contrary, AnnaLee Saxenian, of the University of California at Berkeley, has shown that Indian and Chinese entrepreneurs founded or co-founded roughly 30 percent of all Silicon Valley startups in the late 1990s. But these fundamentally creative people had to travel to Silicon Valley and be absorbed into its innovative ecosystem before their ideas became economically viable. Such ecosystems matter, and there aren't many of them.

Map C (omitted)—which makes use of data from both the World Intellectual Property Organization and the U.S. Patent and Trademark

1. Peaks, Hills, and Valleys

When looked at through the lens of economic production, many cities with large populations are diminished and some nearly vanish. Three sorts of places make up the modern economic landscape. First are the cities that generate innovations. These are the tallest peaks; they have the capacity to attract global talent and create new products and industries. They are few in number, and difficult to topple. Second are the economic "hills"—places that manufacture the world's established goods, take its calls, and support its innovation engines. These hills can rise and fall quickly; they are prosperous but insecure. Some, like Dublin and Seoul, are growing into innovative, wealthy peaks; others are declining, eroded by high labor costs and a lack of enduring competitive advantage. Finally there are the vast valleys—places with little connection to the global economy and few immediate prospects.

2. The Geography of Innovation

Commercial innovation and scientific advance are both highly concentrated—but not always in the some places. Several cities in East Asia—particularly in Japan—are home to prolific business innovation but still depend disproportionately on scientific breakthroughs made elsewhere. Likewise, some cities excel in scientific research but not in commercial adaptation. The few places that do both well are very strongly positioned in the global economy. These regions have little to fear, and much to gain, from continuing globalization.

A. Population

Urban areas house half of all the worlds people, and continue to grow in both rich and poor countries.

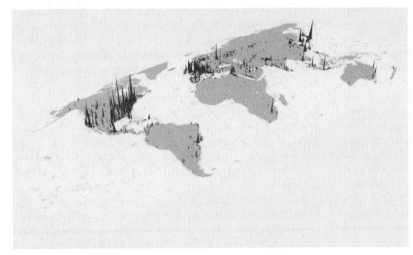

Map reprinted from the Atlantic Monthly, October 2005, "The World Is Spiky," by Richard Florida. Map by Tim Gulden. Data Source: Center for International Earth Science Information Network, Columbia University; and Centro Internacional de Agricultura Tropical

B. Light Emissions

Economic activity—roughly estimated here using light-emissions data—is remarkably concentrated. Many cities, despite their large populations, barely register.

Map reprinted from the Atlantic Monthly, October 2005, "The World Is Spiky," by Richard Florida, Map by Tim Gulden. Source: U.S. Defense Meteorological Satellite Program.

C. Patents

Just a few places produce most of the world's innovations. Innovation remains difficult without a critical mass of financiers, entrepreneurs, and scientists, often nourished by world-class universities and flexible corporations.

D. Scientific Citations

The worlds most prolific and influential scientific researchers overwhelmingly reside in U.S. and European cities.

Office—shows a world composed of innovation peaks and valleys. Tokyo, Seoul, New York, and San Francisco remain the front-runners in the patenting competition. Boston, Seattle, Austin, Toronto, Vancouver, Berlin, Stockholm, Helsinki, London, Osaka, Taipei, and Sydney also stand out.

Map D (omitted) shows the residence of the 1,200 most heavily cited scientists in leading fields. Scientific advance is even more concentrated than patent production. Most occurs not just in a handful of countries but in a handful of cities—primarily in the United States and Europe. Chinese and Indian cities do not even register. As far as global innovation is concerned, perhaps a few dozen places worldwide really compete at the cutting edge.

Concentrations of creative and talented people are particularly important for innovation, according to the Nobel Prize-winning economist Robert Lucas. Ideas flow more freely, are honed more sharply, and can be put into practice more quickly when large numbers of innovators, implementers, and financial backers are in constant contact with one another, both in and out of the office. Creative people cluster not simply because they like to be around one another or they prefer cosmopolitan centers with lots of amenities, though both those things count. They and their companies also cluster because of the powerful productivity advantages, economies of scale, and knowledge spillovers such density brings.

So although one might not *have* to emigrate to innovate, it certainly appears that innovation, economic growth, and prosperity occur in those places that attract a critical mass of top creative talent. Because globalization has increased the returns to innovation, by allowing innovative products and services to quickly reach consumers worldwide, it has strengthened the lure that innovation centers hold for our planet's best and brightest, reinforcing the spikiness of wealth and economic production.

The main difference between now and even a couple of decades ago is not that the world has become flatter but that the world's peaks have become slightly more dispersed—and that the world's hills, the industrial and service centers that produce mature products and support innovation centers, have proliferated and shifted. For the better part of the twentieth century the United States claimed the lion's share of the global economy's innovation peaks, leaving a few outposts in Europe and Japan. But America has since lost some of those peaks, as such industrial-age powerhouses as Pittsburgh, St. Louis, and Cleveland have eroded. At the same time, a number of regions in Europe, Scandinavia, Canada, and the Pacific Rim have moved up.

The world today looks flat to some because the economic and social distances between peaks worldwide have gotten smaller. Connection between peaks has been strengthened by the easy mobility of the global creative class—about 150 million people worldwide. They participate in a global technology system and a global labor market that allow them to migrate freely among the world's leading cities. In a Brookings Institution study the demographer Robert Lang and the world-cities expert Peter Taylor identify a relatively small group of leading city-regions—London, New York, Paris, Tokyo, Hong Kong, Singapore, Chicago, Los Angeles, and San Francisco among them—that are strongly connected to one another.

But Lang and Taylor also identify a much larger group of city-regions that are far more locally oriented. People in spiky places are often more connected to one another, even from half a world away, than they are to people and places in their veritable back yards.

The flat-world theory is not completely misguided. It is a welcome supplement to the widely accepted view (illustrated by the Live 8 concerts and Bono's forays into Africa, by the writings of Jeffrey Sachs and the UN Millennium project) that the growing divide between rich and poor countries is the fundamental feature of the world economy. Friedman's theory more accurately depicts a developing world with capabilities that translate into economic development. In his view, for example, the emerging economies of India and China combine cost advantages, high-tech skills, and entrepreneurial energy, enabling those countries to compete effectively for industries and jobs. The tensions set in motion as the playing field is leveled affect mainly the advanced countries, which see not only manufacturing work but also higher-end jobs, in fields such as software development and financial services, increasingly threatened by off-shoring.

But the flat-world theory blinds us to far more insidious tensions among the world's growing peaks, sinking valleys, and shifting hills. The innovative, talent-attracting "have" regions seem increasingly remote from the talent-exporting "have-not" regions. Second-tier cities, from Detroit and Wolfsburg to Nagoya and Mexico City, are entering an escalating and potentially devastating competition for jobs, talent, and investment. And inequality is growing across the world and within countries.

This is far more harrowing than the flat world Friedman describes, and a good deal more treacherous than the old rich-poor divide, We see its effects in the political backlash against globalization in the advanced world. The recent rejection of the EU constitution by the French, for example, resulted in large part from high rates of "no" votes in suburban and rural quarters, which understandably fear globalization and integration.

But spiky globalization also wreaks havoc on poorer places. China is seeing enormous concentrations of talent and innovation in centers such as Shanghai, Shenzhen, and Beijing, all of which are a world apart from its vast, impoverished rural areas. According to detailed polling by Richard Burkholder, of Gallup, average household incomes in urban China are now triple those in rural regions, and they've grown more than three times as fast since 1999; perhaps as a result, urban and rural Chinese now have very different, often conflicting political and lifestyle values. India is growing even more divided, as Bangalore, Hyderabad, and parts of New Delhi and Bombay pull away from the rest of that enormous country, creating destabilizing political tensions. Economic and demographic forces are sorting people around the world into geographically clustered "tribes" so different (and often mutually antagonistic) as to create a somewhat Hobbesian vision.

We are thus confronted with a difficult predicament. Economic progress requires that the peaks grow stronger and taller. But such growth will exacerbate economic and social disparities, fomenting political reactions that could threaten further innovation and economic progress. Managing the disparities between peaks and valleys worldwide—raising the valleys without shearing off the peaks—will be among the top political challenges of the coming decades.

RICHARD FLORIDA, the author of *The Flight of the Creative Class,* is the Hirst Professor of Public Policy at George Mason University.

Hurricane Hot Spots
Most Vulnerable Cities

Forecasts of another active Atlantic hurricane season raise the possibility of a Katrina-like catastrophe elsewhere in the USA. A look at potential big-city disasters.

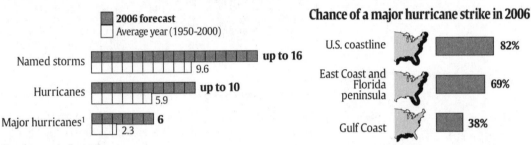

2006 forecast
Average year (1950-2000)

Named storms — **up to 16** / 9.6

Hurricanes — **up to 10** / 5.9

Major hurricanes[1] — **6** / 2.3

Chance of a major hurricane strike in 2006

U.S. coastline — **82%**

East Coast and Florida peninsula — **69%**

Gulf Coast — **38%**

Hurricane wind speeds:
Category 1 - 74-95 mph; **Category 2** - 96-110 mph; **Category 3** - 111-130 mph; **Category 4** - 131-155 mph; **Category 5** - over 155 mph
1 - Major hurricanes are Category 3 and higher

Areas with the most to lose

Steve Lyons, tropical weather expert at The Weather Channel, says the following metropolitan areas would be particularly vulnerable if hit by a major hurricane. Risk maps were provided by AIR Worldwide.

New York City

New York is perhaps the least-susceptible coastal city for a hurricane strike, but it is by far the metropolis with the most infrastructure at risk. Although a direct hit is a long shot, Lyons says, there is precedent. During the last century, six storms slammed the New York coastline. In 1893, a Category 2 storm swamped the city and wiped out Hog Island—a mile-long getaway for politicians and business elite.

Previous major hurricanes in the area: Carol (1954), Donna (1960), Gloria (1985)

Challenges if a major storm hits:

- More than 2.5 million people could be forced to evacuate lower elevations.
- Storm surge could flood subways, overrun major airports and submerge much of western Long Island and Lower Manhattan.
- Up to 800,000 displaced people could require shelter.
- AIR Worldwide, a risk-modeling company, estimates insured losses at $100 billion if a Category 4 hurricane hits the New York-New Jersey region.

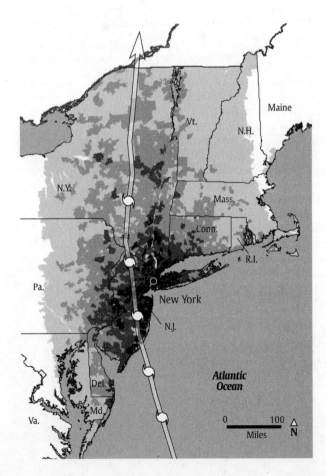

Houston-Galveston

Galveston built a 7-mile, 17-foot-tall sea wall after a hurricane in 1900 killed more than 6,000 people. Neighboring Houston has not done as much to protect itself from a direct hit.

Previous major hurricanes in the area: Audrey (1957), Carla (1961), Alicia (1983), Rita (2005)

The challenges:

- Major infrastructure built along Galveston Bay is vulnerable to storm surge.
- An arduous evacuation before Hurricane Rita demonstrated the difficulty of getting people out of harm's way.
- Houston, a low-lying city, would have extensive flood damage. Frequent summer rains likely would destroy contents of homes with roof damage from winds.
- AIR Worldwide estimates a direct hit by a Category 5 hurricane would result in insured losses exceeding $60 billion.

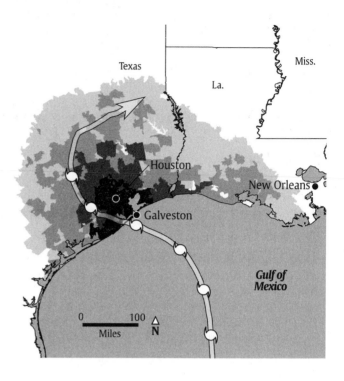

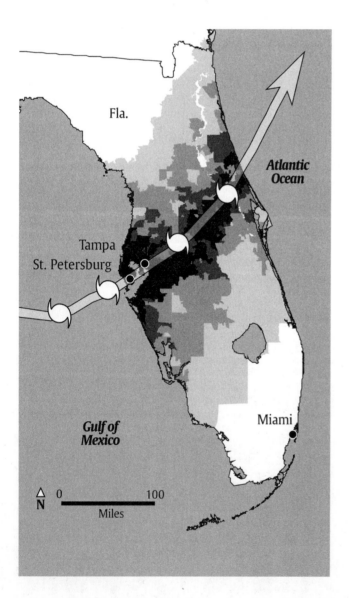

New Orleans

Unfinished levee repairs and improvements make this city on the Mississippi River more vulnerable to storm surge than pre-Katrina, Lyons says.

Tampa-St. Petersburg

Lyons calls Tampa "the untested city" because of recent development in low-lying areas. Hurricane Frances in 2004 caused extensive damage and left nearly 100,000 people without power, but it approached from the east and did not bring a major storm surge.

Previous major hurricanes in the area: Easy (1950) and Charley (2004)

The challenges:

- The shape and depth of Tampa Bay make it prone to surges.
- Floodwaters would cover many of the expensive waterfront homes.
- Evacuation over bridges or city streets would be time-consuming. Evacuees might not be able to escape the storm inland or find housing during tourist season.
- AIR Worldwide estimates a direct hit by a Category 5 hurricane would result in insured losses exceeding $70 billion.

Estimated insured losses in millions of dollars (by ZIP code)

More than 50   11 to 50   1 to 10   Below 1

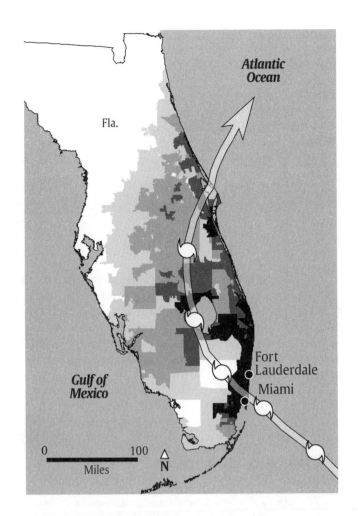

Miami-Fort Lauderdale

Miami's population density and massive infrastructure make it the most vulnerable area in the nation's most hurricane-prone state.

Previous major hurricanes in the are: King (1950), Betsy (1965), Andrew (1992)

The challenges:

- In 2005, Hurricane Wilma caused $10 billion in damage when it hit the Miami area with only Category 1 winds.
- The tightly packed metro area—5.4 million people—would make an evacuation difficult.
- Surge should not be a problem, Lyons says, because of the depth of coastal waters and the Bahamas.
- AIR Worldwide estimates a direct hit by a Category 5 hurricane would result in insured losses exceeding $130 billion.

What's at stake

The estimated insured value of commercial and residential property in Gulf and Atlantic coastal areas exceeds $7 trillion, according to AIR Worldwide.

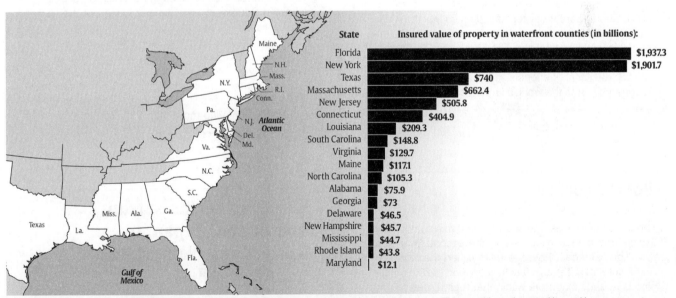

State	Insured value of property in waterfront counties (in billions):
Florida	$1,937.3
New York	$1,901.7
Texas	$740
Massachusetts	$662.4
New Jersey	$505.8
Connecticut	$404.9
Louisiana	$209.3
South Carolina	$148.8
Virginia	$129.7
Maine	$117.1
North Carolina	$105.3
Alabama	$75.9
Georgia	$73
Delaware	$46.5
New Hampshire	$45.7
Mississippi	$44.7
Rhode Island	$43.8
Maryland	$12.1

Note: Actual losses would include insured and uninsured property as well as areas repaired by governments. Those could greatly exceed insured losses.

AIR Worldwide: The Weather Channel: Department of Atmospheric Science, Colorado State University: National Hurricane Center: N.Y. Office of Emergency Management: Army Corps of Engineers Research by Chris Fruitrich and Bob Swanson, graphic by Dave Merrill. USA TODAY

Sea Change
The Transformation of the Arctic

SCOTT BORGERSON

Before our eyes, the Arctic is changing from an impenetrable wasteland into an oceanic crossroads. The polar ice cap has lost up to half its thickness nearfgfgthe North Pole in just the past six years and may have passed a tipping point; it is now shrinking at more than three times the rate predicted by the Intergovernmental Panel on Climate Change only four years ago. At the current pace, the Arctic may well be ice-free in summer by 2013.

The opening of a new waterway between the Atlantic and Pacific oceans is akin in historic significance to the opening of the Suez Canal, in 1869, or its Panamanian cousin, in 1914. With this sea change will come the rise and fall of international seaports, newfound access to nearly a quarter of the world's remaining undiscovered oil and gas reserves, and a recalibration of geostrategic power.

Drill, Baby, Drill?

A U.S. Geological Survey report estimates that the Arctic holds 90 billion barrels of undiscovered oil and 1,670 trillion cubic feet of natural gas. The Shtokman Field alone, in Russian waters, contains enough natural gas to power the U.S. electrical grid for six years, and may soon help Gazprom become the world's largest company.

The New Entrepôts

Singapore's location, amid key shipping lanes from East Asia to Europe, has enabled the country to become the richest in Southeast Asia. As sea lanes like the Bering and Davis straits become busier, port towns like tiny Dutch Harbor, Alaska (population 4,000), and Hammerfest, Norway (population 9,000), are likely to grow from out-of-the-way fishing depots into key shipping hubs. Russia recently committed $7 billion to port development in Murmansk. Places like Singapore or Panama (which is currently investing more than $5 billion to expand its canal) may see trade disappear from their doorsteps.

Polar Express

The fabled Northwest Passage opened this summer for the second time in history—and the second year in a row. The Northeast Passage (also called the Northern Sea Route) over Eurasia first fully opened in 2005; shipping is already extensive within that region, particularly in the Barents Sea. Yet both routes, sought by ancient mariners, are likely to be used for only a few years. By 2025, if not before, most ships in the Arctic will likely sail straight over the pole, avoiding coastal-state jurisdictions and shaving still more miles off their journeys. Much of the world's international shipping will reorient itself as a result.

Shipping Distance from Yokohama, Japan, to Rotterdam, Netherlands (nautical miles)

Over the North Pole	**5,618 miles**
Through the Suez Canal	11,209 miles
Through the Panama Canal	12,250 miles
Around the Cape of Good Hope	14,735 miles

(Financial) Freedom

Retreating ice is revealing up to 31 billion barrels of oil and natural gas off Greenland's eastern coast, plus signs of enormous mineral deposits—gold, diamonds, zinc, and more—on land. Keen to establish their rights to these resources, 57,000 Greenlanders (most of them Inuit), whose territory has been protected by Denmark since 1721, will vote this month on a referendum for self-rule. Greenland will probably become the first country born from climate change.

Snow Forts

To defend its claims in the Arctic, Canada plans to build a deepwater naval port in Nanisivik and a new cold-weather-combat training center in Resolute Bay, while also expanding satellite surveillance. Other countries are rattling sabers in the region. Russia, for example, has resumed strategic-bomber flights over the Arctic and last summer dispatched two military vessels to the disputed waters off the Svalbard Islands.

SCOTT BORGERSON is the Visiting Fellow for Ocean Governance at the Council on Foreign Relations.

Shaping the World to Illustrate Inequalities in Health

Danny Dorling and Anna Barford

Visualizing inequalities in health at the world scale is not easily achieved from tables of mortality rates. Maps that show rates using a colour scale often are less informative than many map-readers realize. For instance, a country with a very small land area receives less attention, whereas a large, sparsely populated area on a map is more obvious. Furthermore, unlike our visual ability to compare the lengths of bars in a chart, we do not have a natural aptitude for translating different colours or shades to the magnitudes they represent. Here we introduce another approach to mapping the world that can be useful for illustrating inequalities in health.

Where do you think most infants in the world are born, where do most die and how have these measures changed since 1970? A map of birth rates would not help you much, unless you had the kind of memory that could associate several hundred areas with counts of their populations of young women, and had the ability to perform some quick mental arithmetic of rate reciprocals. Nor would maps of death rates help much in answering these questions. Seeing the world shaped by how many babies are born in a year is a more reliable and rapid way of communicating these numbers (Figure 1).

This figure was created using software derived from that which is freely available online.[1] The software changes the sizes of countries to represent the proportion of all children worldwide who were born there. This is done by equalizing the densities of a measure such that a country that is physically quite small but with many births increases in size (e.g. Guatemala), while somewhere with a large area but few births shrinks (e.g. Australia). Each birth is allocated the same amount of space, and thus country borders are stretched and crumpled around these adjusted areas.[2,3]

Very few people can identify most countries of the world on an unlabelled conventional world map. Country identification can be even more difficult on the cartogram that is Figure 1, given the distortion from the world shape that we are used to seeing. However, this particular cartogram does have the advantage of inviting the reader to focus on identifying those countries in which most of humanity is born. And as the topology of the planet is preserved and the shape often not too distorted, this is not too difficult a task,[4] and is far easier than imagining

rates from colours. The shades in Figure 1 are there to allow the matching of countries between the different maps in this paper. Compare Figure 1 to Figure 2—countries that are larger in the former have lower rates of infant death, and vice versa.

Figure 2 is shaded and made identically to Figure 1, except now the countries are sized by the numbers of infants who died in each country in their first year of life in the year 2002. By comparing the two maps you gain an impression of not just where rates are higher or lower than the world average, but also of how many infants are affected by these inequalities. In both maps, the total areas of all countries are identical. Our ability to gauge and compare areas is not great; it is worse than our general visual aptitude for comparing lengths. But if, when comparing these two maps, you think you are benefiting from this despite almost certainly already being aware of these inequalities, then these maps should be useful for those less familiar with this subject.

There are of course limitations of the data, as well as of the technique. In only a minority of countries are all births and deaths registered, so these maps mostly show estimates. However, to varying degrees that is true of all such data. Because the quality of data varies between countries, there is a danger of these images inviting flawed comparisons to be made. It is not possible to show "missing data" on these maps; giving them zero area would imply no births or deaths. That would mean that countries experiencing some of the worst of times might be ignored, as they often are when missing from official tabulations. The disadvantage of estimating data is that the numbers for countries such as Afghanistan, Iraq and Somalia may not be quite as accurate as others. However, the degree of inaccuracy involved with these figures is probably less than the error in your visual estimation. Just how circle-shaped a country is can affect how large you think it is.

These maps use countries as their unit of analysis, thus obscuring sub-national variations which themselves are sometimes greater than the international variations. To use cartograms at a smaller scale would require finer-grained data, which is subject to the same limitations as country-level data. Also, there are inevitable variations in data collection or estimation, definitions used and varying political motives behind particular

Figure 1   Worldmapper map 3: total births

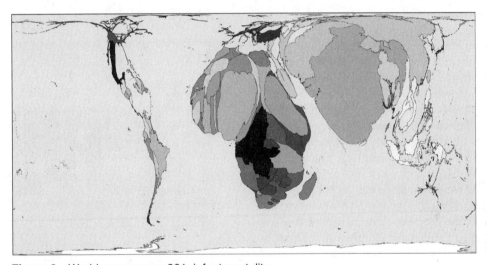

Figure 2   Worldmapper map 261: infant mortality

numerical descriptions of health in different places. As is the case with national data, those areas that are the most disrupted are often where we know least about what is going on. Nevertheless, for policy and planning such maps could be an effective tool to draw attention to what is happening where, and to possibly guide resource allocation. An example of the political use of these maps was that they were used in the 2006 International Monetary Fund discussions about vote redistribution.[5]

Map area can be used to show something that requires urgent action, such as high infant deaths, but it can also show successes such as large decreases in those deaths. If we can map something that is happening, we can also map its inverse—where it happens less over time, it is also possible to show these changes. Figure 3 illustrates the extent of improvement in infant mortality over the past three decades in terms of the number of infants who survived until their first birthdays in 2002 who would have died had the infant mortality rates of 1970 continued. Reading Figure 3 in conjunction with Figure 2, we can develop a sense of where improvements have occurred in the context of where most infant deaths still happen.

No country experienced an increased rate of infant mortality between 1970 and 2002, but if one had there would be no difference in its appearance on Figure 3 from if the mortality rate had remained the same: it would still have zero area. Maps of change over time are limited because these two-dimensional images cannot be easily used to express negative area—so increases and decreases cannot be drawn on the same map. Comparing between maps showing the inverse of the same variable or showing different variables can help us to see certain patterns. Yet there are other patterns that are much more clearly expressed by different means. Figure 4 shows the percentage change in infant mortality by region—it provides a clear depiction of how closely a region's position on the Human Development Index (by which the regions are ordered) relates to the extent of improvements to infant mortality. Read in conjunction with the other figures, we can see changes in the rates as well as where these improvements are happening, and get a sense of the distribution of how and where children are living longer.

In reading these maps it is worth considering what is being shown and not shown, and what insights you are gaining.

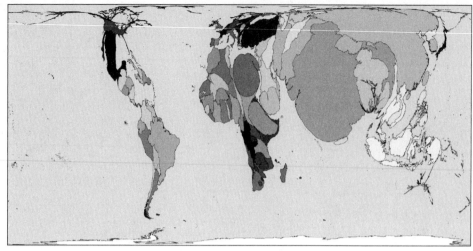

Figure 3 Worldmapper map 262: infant mortality change

Keeping in mind the international variability in data quality and availability, do you now have an improved sense of the worldwide distribution of births, of infant mortality and of the extent of improvements in infant mortality? In doing this, we can see that some of the poorest countries in the world, where there were the most infant deaths in 2002, have seen far fewer infants die than did at 1970s rates. However, the graph informs us that it is in the richer countries where the largest proportionate improvements in infant mortality rates have occurred; these are territories where there were relatively few deaths to begin with.

The three maps that are shown here form part of series that is freely available at www.worldmapper.org. Other maps in the series include health-care provision, distribution of disease, wealth, poverty, trade and pollution.[6] The website is about to launch a series of maps of the distribution of different causes of death based on World Health Organization Global Burden of Disease data. Each map is accompanied by datasheets, technical notes and a downloadable poster. Hopefully this website allows viewers to not only see one country in its global context, but also to see one element of our lives in relation to another.

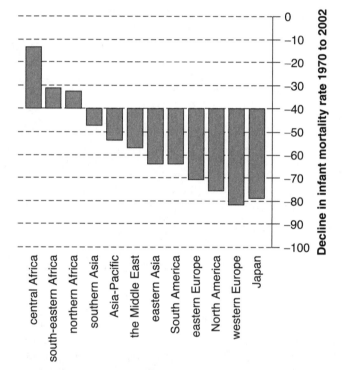

Figure 4 Decline in rate of infant mortality, 1970–2002

References

1. MT Gastner, MEJ Newman. Diffusion-based method for producing density equalizing maps. *Proc Natl Acad Sci USA* 2004; 101: 7499–504.
2. A Barford, D Dorling. The world: a different view. *Natl Med J India* 2006; 19:234–5.
3. Dorling D. Anamorphosis: the geography of physicians and mortality. *International Journal of Epidemiology* 2007. doi: 10.1093/ije/dym017. Available as pre-print from: http://ije.oxfordjournals.org/cgi/content/full/dym017v110.1093/ije/dym017
4. D Dorling. Worldmapper: the human anatomy of a small planet. *PLoS Med* 2007; 4: e1-.
5. *The map of votes in the International Monetary Fund circulated to members of the African Caucus.* Available at: http://www. worldmapper.org/display.php?selected=365 www.sasi.group .shef.ac.uk/worldmapper/articles/IMF_voting_share.pdf
6. D Dorling, A Barford, M Newman. Worldmapper: the world as you have never seen it before, I. *IEEE Trans Vis Comput Graph* 2006: 757–64.

DANNY DORLING, Social and Spatial Inequalities Group; **ANNA BARFORD** Department of Geography, University of Sheffield, England.

Acknowledgements—Other researchers working on the Worldmapper project are John Pritchard of the University of Sheffield and Mark Newman of the University of Michigan. The Leverhulme Trust provided financial support for this work.

Deaths Outnumber Births in Third of Counties

"It's Like a Final Spiral" in Some Areas.

HAYA EL NASSER

A Catholic priest in rural Kansas handles four times as many funerals as baptisms.

Deaths in one Florida county outnumber births so dramatically that if people hadn't moved in from other places, the population would have plummeted.

In North Dakota's Sheridan County, many young people have left, and the aging residents who remain are dying. The only practicing lawyer left in the county is the prosecutor.

Almost a third of U.S. counties are experiencing more deaths than births, according to research by Kenneth Johnson, senior demographer at the University of New Hampshire's Carsey Institute.

Their numbers hit a record 988 in 2002, more than twice as many as in 1990. They've stayed above 900 ever since.

This natural decrease—deaths outnumbering births—shows that the USA's expansion hasn't reached every corner of the country. The aging of the population, however, has a profound effect nationwide.

The fact that 909 counties in 2006 needed more coffins than cradles in a nation that has grown to more than 300 million and has a higher fertility rate than most of the developed world surprised even Johnson, who has analyzed the trends for three decades.

"What does it do to a community to have the median age be 55 and for more people to be dying than to be born?" he asks. "It's not the kind of community America is used to. . . . When natural decrease occurs and there's no in-migration, it's like a final spiral."

Johnson found that these naturally declining counties fit into one of two categories:

- **Shrinking counties.** They're in rural areas far from metropolitan centers. Many are in parts of the Great Plains and East Texas. They're counties that have been abandoned by young people for at least one generation. Some have experienced natural decreases 35 years in a row.

"It's gone so far that you have these counties facing a double whammy," says Calvin Beale, senior demographer at the Agriculture Department. "They continue to have people moving out, and there are more people available to die than there are to have children."

- **Retirement counties.** They still gain population and often thrive, but they have more deaths than births because they have so many seniors.

"Counties that have in-movement of older people typically have in-movement of younger people as well," Beale says.

They attract immigrant labor to fill jobs in restaurants, hospitals, retirement communities and other support services.

"In some areas, the influx of young Hispanics is offsetting losses in the white population, most of it from natural decrease," Johnson says.

Some Counties Emptying

Walter Lipp is the local prosecutor in Sheridan County, N.D., and the only lawyer in this wheat-growing area in the center of the state. The county of 1,408 lost one-sixth of its population in six years and had twice as many deaths as births during that time.

"The other lawyer who was here died last year," Lipp says. "And he was in a nursing home for four or five years."

The county prosecutor's job is not even full time. Lipp is in private practice the rest of the time.

"The forecast is the same it's been for the last 30 or 40 years," he says. "The farms get bigger and bigger, and the population goes down. Kids move away. There are limited jobs. There's really no relief in sight."

The median age in Sheridan County is 52.9. The population of its main town, McClusky, slipped from 415 in 2000 to 337 in 2006. There are seven churches in town,

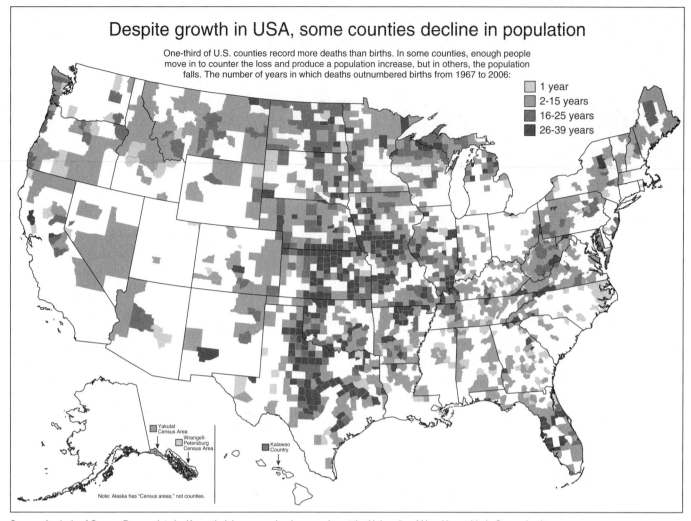

Despite growth in USA, some counties decline in population

One-third of U.S. counties record more deaths than births. In some counties, enough people move in to counter the loss and produce a population increase, but in others, the population falls. The number of years in which deaths outnumbered births from 1967 to 2006:

- 1 year
- 2-15 years
- 16-25 years
- 26-39 years

Yakutat Census Area

Wrangell-Petersburg Census Area

Kalawao Country

Note: Alaska has "Census areas," not counties.

Source: Analysis of Census Bureau data by Kenneth Johnson, senior demographer at the University of New Hampshire's Carsey Institute

but their ministers also serve churches in two or three other towns. "Everything is consolidating," Lipp says.

In western Kansas' Ness County, it has been more than two years since the town of Utica's school district dissolved, requiring youngsters to ride the bus to a neighboring town for education. Since 2000, Ness' population dropped 15% to fewer than 3,000 people and its median age soared from 43.9 to 49.1. In 2006, 34 people died and 24 were born.

The New Homestead Act, introduced by Sen. Byron Dorgan, D-N.D., in April, is intended to stem the decline in such counties. Dorgan's proposal would provide tax credits for home purchases and small businesses and repay up to 50% of college loans, among other things.

Although the Senate has approved various parts of the bill, none has been enacted, according to Barry Piatt, Dorgan's spokesman.

"In some of these areas, I don't think government policy could stop the decline," says Johnson, the demographer who researched the trend. "There were a lot of counties in the Great Plains that had more people than they should

have had. Agriculture has changed. You don't need as many people working the land. . . . You can think of it as sort of a triage."

Growth Despite Deaths

Deaths may outnumber births in Sarasota County on Florida's Gulf Coast, but the region is growing.

It began attracting thousands of retirees in the 1990s. It also appealed to working-age baby boomers because the area, south of St. Petersburg, is affordable, says Warren Richardson, spokesman for the county. More recently, Ukrainian immigrants have settled around North Port in the southern part of the county, he says.

Just more than 3,000 were born in the county in 2006, not enough to offset the 4,940 deaths there that year. Despite that, the population grew 1.2% from 2005 to 2006, and more than 13% since the beginning of the decade to almost 370,000.

"That the population is aging is something we're very much aware of," Richardson says. "We don't have specific policies in place. . . . But we're getting our share of the younger crowd."

No one expects counties that are experiencing natural decreases to be wiped off the map.

"They won't empty out altogether," Beale says. "There's land there, and people are occupying the land, even if it's with very large ranches and very large farms."

Johnson finds some irony in his research. "This is a country focused on growth, on the population getting bigger, on expansion and sprawl," he says. "We have near-record numbers of births right now, immigrants are flooding in. . . . So we have endless arguments about immigrants and smart growth while hundreds of counties are essentially being depopulated with no one noticing."

How Much Is Your Vote Worth?

Sarah K. Cowan, Stephen Doyle, and Drew Heffron

"The conception of political equality from the Declaration of Independence, to Lincoln's Gettysburg Address, to the Fifteenth, Seventeenth and Nineteenth Amendments can mean only one thing—one person, one vote," the Supreme Court ruled almost a half-century ago. Yet the framers of the Constitution made this aspiration impossible, then and now.

Under the Constitution, electoral votes are apportioned to states according to the total number of senators and representatives from each state. So even the smallest states, regardless of their population, get at least three electoral votes.

But there is a second, less obvious distortion to the "one person, one vote" principle. Seats in the House of Representatives are apportioned according to the number of residents in a given state, not the number of eligible voters. And many residents—children, noncitizens and, in many states, prisoners and felons—do not have the right to vote.

In House races, 10 eligible voters in California, a state with many residents who cannot vote, represent 16 people in the voting booth. In New York and New Jersey, 10 enfranchised residents stand for themselves and five others. (And given that only 60 percent of eligible voters turn out at the polls, the actual

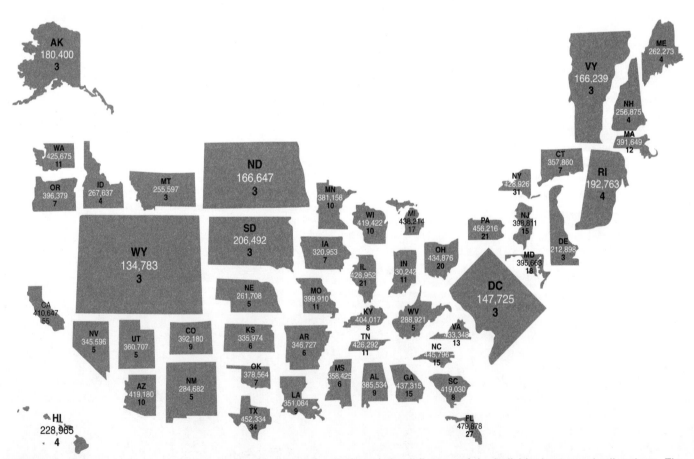

Figure 1 *This map shows each state re-sized in proportion to the relative influence of the individual voters who live there. The numbers indicate the total delegates to the Electoral College from each state, and how many eligible voters a single delegate from each state represents.*

Source: The Uniter States election project at George Mason University.

figures are even starker.) Of all the states, Vermont comes the closest to the one person, one vote standard. Ten Vermont residents represent 12 people.

In the Electoral College, the combined effect of these two distortions is a mockery of the principle of "one person, one vote." While each of Florida's 27 electoral delegates represents almost 480,000 eligible voters, each of the three delegates from Wyoming represents only 135,000 eligible voters. That makes a voter casting a presidential ballot in Wyoming three and a half times more influential than a voter in Florida.

This system, along with the winner-take-all practice used to allocate most states' electoral votes, creates the potential for an absurd outcome. In the unlikely event that all 213 million eligible voters cast ballots, either John McCain or Barack Obama could win enough states to capture the White House with only 47.8 million strategically located votes. The presidency could be won with just 22 percent of the electorate's support, only 16 percent of the entire population's.

Sarah K. Cowan is a graduate student in sociology and demography at the University of California, Berkeley. **Stephen Doyle** and **Drew Heffron** are graphic designers.

Mapping

Fortune Teller

For magazine readers of the 1940s, architect-turned-artist Richard Harrison mapped their world in a bold new way.

ANN DE FOREST

"Three approaches to the U.S.," reads the headline. Underneath, printed in vivid color on a page of the old magazine, are three maps, three rounded edges of the globe, overlapping, like an illustration of a planet rising. In each, the land spreads out tan and yellow against the blue-green water. It almost seems that you're flying over this contoured terrain, about to swoop down for a landing. But what terrain is it? The headline says it's the United States, but these maps don't look like any America you know. Where is that familiar, iconic national outline? Where is Maine, waving in the upper right-hand corner? And where is California, bending forward like a ship's prow on the left?

The man who made these maps, Richard Edes Harrison, cartographer for *Fortune* for more than 20 years, specialized in such unorthodox presentations. He sought to, in his own words, "jolt . . . [readers] . . . with a new and refreshing viewpoint." So, on a Harrison map, north isn't always up, the earth isn't always flattened on the page and perspective—the point from which one views the planet—shifts, depending on what information he's aiming to convey. Says Joanne Perry, maps librarian and head of cartographic services at Pennsylvania State University. "He was trying to put out maps that showed truth, not convention."

In September 1940, the truth these particular maps revealed was far more ominous than refreshing. More than a year before the Japanese attack on Pearl Harbor, *Fortune* featured the images as part of an *Atlas for the U.S. Citizen:* 11 maps designed to show Americans they weren't as isolated from events in Europe and Pacific as they might like to believe. That swooping perspective, so breathtaking today, is actually the enemy's-eye view. Approached by air from Berlin, Tokyo or even Caracas, Venezuela, the United States suddenly looks vulnerable, with Detroit and Chicago tempting targets for the Germans, and Seattle and San Francisco poking perilously in Tokyo's flight path. And "if an enemy . . . ," a caption warns, "should ever establish himself on the northern shore of South America or in the mazes of the West Indies," the Gulf Coast, hub of industry, becomes America's "soft belly."

Harrison, architect by training, artist by inclination, came to mapmaking accidentally (though he'd done some scientific illustrations for pay before studying architecture). Like many Americans in 1932, he needed work. A friend recommended him for an assignment to *Time* to make a map projecting the outcome of the looming presidential election. In 1936, he was kicked upstairs to another magazine in the Henry Luce empire, *Fortune,* where as part of its crack graphics team, he applied his visual skills to explicating the macro—charting the worldwide holdings of General Motors—and the micro—diagramming the inner workings of a gas mask.

World War II proved the ideal subject for his talents. This was a new kind of war, a truly global war, fought in the skies as well as on land and at sea. It was a war that demanded entirely new maps, new ways of seeing the world. Here, Harrison's lack of formal cartographic training was an advantage. Not bound by convention, he could make maps that opened readers' eyes to the realities of global geography in an aviation age.

"He explicitly saw himself as speaking to the general public, while he saw cartographers as failing in this regard," says Susan Schulten, a history professor at the University of Denver and author of *The Geographical Imagination in America, 1880–1950.* She interviewed the "witty, subversive" carto-journalist, then 90 years old, shortly before his death in 1993. Certainly, the wide range of maps he made for *Fortune* in the 1940s show him reveling in his amateur status. He tweaks professional geographers for their staunch commitment to the Mercator projection, the standard flat map of the world ("a dangerous map for world strategy," reads one caption). And he always finds space to explain, through clever illustrations and entertaining examples, exactly why he chose a particular style of mapmaking for a particular purpose. Never condescending, Harrison takes the tone of a fellow layman, and enthusiastic

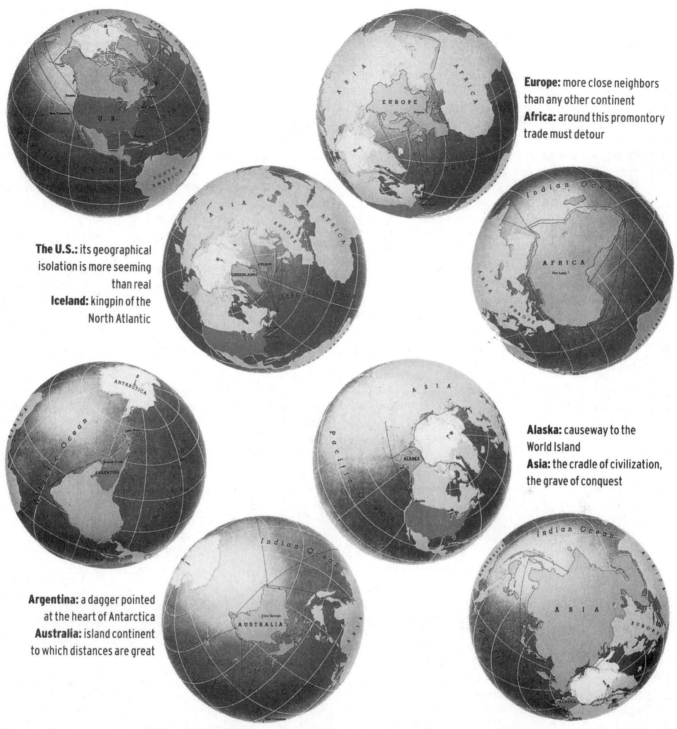

Europe: more close neighbors than any other continent
Africa: around this promontory trade must detour

The U.S.: its geographical isolation is more seeming than real
Iceland: kingpin of the North Atlantic

Alaska: causeway to the World Island
Asia: the cradle of civilization, the grave of conquest

Argentina: a dagger pointed at the heart of Antarctica
Australia: island continent to which distances are great

Planet Life Harrison's "Eight Views of the World" was created for *Look at the World*. The captions are from the original. Used with permission of Ross & Harrison III.

amateur, who just happened to discover what terms like "azimuthal equidistant" meant himself.

His map of "The World Divided," published in August 1941, centers on the North Pole, with the rest of the world a great circle around it. (In a typically charming sidebar, Harrison explains the principle behind this map with two cartoons of a dancer. In the first, she stands at rest, with her skirt as the globe; in the next, she's twirling, with the globe having risen and flattened to a disk.) This world, "divided" into Axis nations, Anti-Axis and various positions in between by color and shading, is also inextricably united: The U.S.S.R. and Alaska touch on this map, while the Aleutian Islands are part of one long, necklace-like chain that extends through Japan and down to the Philippines.

This view was prescient, of course. In March 1942, a nearly identical "polar equidistant" map appeared in *Fortune,* titled "One World, One War." Indeed, once the United States entered the war, Harrison's maps became integral to *Fortune*'s reports on U.S. action and changing strategic situations worldwide. The sweeping, slightly rounded views gave "a sense of direction, a sense of the movement of the war," says Schulten.

Harrison's work also serves to underscore the advances in air navigation at the time. As Schulten wrote in her book, "The use of a polar route to connect Japan to Alaska effectively transformed the Pacific from a massive body of water protecting the United States into a smallish lake."

In 1942, *Fortune* printed a stunning series of colorful, foldout maps of the Pacific, Atlantic and Arctic arenas. These, says Perry, "make you see why something is happening in the world: Why are our troops in North Africa? . . . Why are your sons and neighbors dying in different places?" His vivid creations were phenomenally popular, reprinted by the military and various airlines and displayed at post offices.

Capitalizing on the cartographer's celebrity, *Fortune* published a book of Harrison's wartime maps in 1944. *Look at the World: The Fortune Atlas for World Strategy* was an instant best seller. Immediately after World War II, Harrison collaborated on numerous books like *Compass of the World: A Symposium on Political Geography* and *Maps and How to Understand Them,* an unusual hybrid textbook/political tract/advertising pamphlet published by Consolidated Vultee Aircraft Corporation promoting "Air Supremacy—For Enduring Peace."

Today, though, *Fortune*'s "celebrated cartographer" is all but forgotten. The Mercator projection—that flat, 433-year-old map designed for an age of navigation, not aviation—has prevailed. "Perhaps his style of maps were really keyed into a time of crisis," Schulten speculates. "He had an advantage in the 1940s. While Rand McNally had to create maps that would last," Harrison, working for a monthly magazine, was making maps that were newsworthy. "He was drawing not just for the war, but for that week of the war."

We all know the fate of yesterday's news. It's too bad, though. Harrison's fresh-eyed perspectives and projections deserve a revival. Crisis or not, when it comes to viewing the world, we can always stand a "jolt . . . with a new and refreshing viewpoint."

Speaking of fortune, *Navigator* is fortunate to have **ANN DE FOREST** as its resident cartographile.

Teaching Note: The U.S. Ethanol Industry with Comments on the Great Plains

Executive Summary

In 1990, the U.S. Congress passed amendments to the Clean Air Act, establishing two programs to reduce automotive pollution by mandating "cleaner" fuel. The Oxygenated Fuels Program was targeted at reducing carbon monoxide emissions, while the Reformulated Gasoline Program was intended to reduce smog-forming emissions. Ethanol and methyl tertiary butyl ether (MTBE) are the two main oxygenates (i.e., additives that increase the oxygen content in fuel) used to meet the requirements of these programs. The implementation of these two programs has stimulated ethanol demand considerably, and as a result ethanol production has nearly doubled since 1990. A large percentage of this increased production is located in the Great Plains and upper Midwest.

Recently, several U.S. states have taken steps to ban MTBE due to contamination of groundwater and alleged adverse health effects. And, legislation is pending in Congress that would phase out the use of MTBE nationally. Even without a complete prohibition against the use of MTBE, trends that are already in place suggest a modest increase in ethanol consumption over the course of this decade. However, if MTBE is banned and the oxygenate requirements are maintained in the Clean Air Act programs, growth in ethanol consumption could be dramatic. Alternatively, the proposals in Congress for a possible nationwide renewable fuels standard for U.S. gasoline could also result in a sizable increase in ethanol consumption. These forces suggest that the consumption of ethanol may well continue to increase substantially in the coming years. The close proximity to corn, the main feedstock for ethanol, may offer an attractive option for Great Plains producers who wish to look for option value added markets for grain production.

The expectations are based on prospects for renewed strength in gasoline prices, thus, ethanol prices are expected to remain at or above historical averages unless significant surplus capacity is built. Combined with expectations that corn prices will remain at or below long-term averages unless a new Farm Bill places significant restrictions on farmers, ethanol processing margins are likely to remain at or above historical averages, though not as strong as margins experienced in 2001. Thus, the current outlook is deemed favorable for adding to capacity over the forecast timeframe.

Introduction

In the United States, ethanol has been described as a "political commodity," due to the importance of federal government policies that stimulate the production and consumption of ethanol and make it price-competitive in fuel markets. The primary federal incentive has been a partial exemption of ethanol-blended fuels from the federal excise tax on motor fuels. However, even with this partial exemption, fuel ethanol production had begun to level off in the late 1980s (see Figure 1).

Then, in 1990, the U.S. Congress passed amendments to the Clean Air Act (referred to as CAA90), establishing two programs to reduce automotive pollution by mandating specifications for "cleaner" fuel. The Oxygenated Fuels Program was targeted at reducing carbon monoxide emissions, while the Reformulated Gasoline Program was intended to reduce smog-forming emissions. Ethanol and methyl tertiary butyl ether (MTBE) are the two main oxygenates (i.e., additives to fuel to increase the oxygen content) used to meet the requirements of these programs.

The CAA90 stimulated ethanol demand considerably, and in response ethanol production has nearly doubled. The economics of ethanol production were particularly strong in 2000 and the first three quarters of 2001 due to a combination of low corn prices and relatively high gasoline prices. As a result, U.S. ethanol production reached a record 1.63 billion gallons in 2000 and will be about 2.3 billion gallons in 2002.

Moreover, several U.S. states have taken steps to ban MTBE, the main competitor against ethanol in the fulfillment of the CAA90 oxygen requirements. These bans have been ordered because of the contamination of groundwater by MTBE in a number of locations, along with allegations

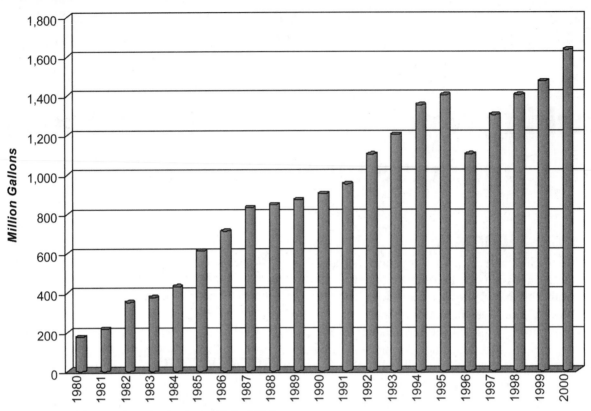

Figure 1 Fuel ethanol production in the United States.
Source: Energy Information Administration and Renewable Fuels Assn.

that MTBE causes human health problems. Of particular note is California, which is by far the largest gas-consuming state in the country. California has implemented its own statewide reformulated gasoline program, though 70% of the state's gasoline consumption is in metropolitan areas that would otherwise have been subject to the federal CAA90 programs. California has banned the use of MTBE after December 31, 2002, and requested a waiver from the federal oxygenate standard (as of this writing the ban was delayed one year). This request was denied by the Environmental Protection Agency (EPA) in the summer of 2001.

As a result of the favorable processing margins and policy developments in the U.S., during 2000 and most of 2001 a number of producer cooperatives and companies are implementing or considering the construction of new ethanol facilities and/or the expansion of existing facilities. The Great Plains region (roughly defined as the plains states west of the Mississippi from North Dakota to Texas), due to attractive economics versus other regions, will garner a large share of this expected increase in production capacity.

Policy Environment
United States Background

The economic and political shocks from the OPEC oil embargoes in 1973 and 1979 generated widespread political pressure in the United States to reduce dependence on imported petroleum. Concerns about exploitation by the petroleum cartel generated a broad range of proposals for new energy sources, including domestic production from renewable resources. These programs involved primarily federal research, including federal support for a wide variety of regional and state projects. The United States Department of Agriculture (USDA), in conjunction with the U.S. Department of Energy (DOE) and the land-grant university system, developed a large number of research projects to utilize grain products and animal/crop wastes, and to test many other approaches. Subsidies for the construction of facilities to produce energy became widespread, including concessional interest rates and grants.

When petroleum costs were high in the 1970s and early 1980s, enthusiasm for these programs was strong, especially in the farm community, which saw them as a new and potentially large source of demand for farm and forest products. However, petroleum costs declined in the mid to late 1980s and 1990s, and products from renewable resources became increasingly less competitive. As a result, production of some forms of renewable energy were abandoned, except where supported by subsidies. For ethanol, program funding and product usage continued mainly through support from the energy research community and the agricultural lobby.

By the late 1980s, the debate over use of renewable fuels had changed significantly. Pressure from environmentalists and others for cleaner air led to requirements for oxygenates to be used in automotive fuels, in order to reduce pollution. In 1990, Congress passed amendments to the Clean Air Act (CAA90), establishing two programs to reduce pollution from automotive emissions. Whereas previous legislative and administrative pollution-control requirements had focused mainly on creating more fuel-efficient cars, the CAA90 focused on the composition of cleaner burning automotive fuel.

Clean Air Act Amendments of 1990

The CAA90 created two programs mandating changes in fuel composition, in order to address two distinct pollution problems: the Oxygenated Fuels (OXY) Program aimed at carbon monoxide, and the Reformulated Gasoline (RFG) Program targeted at smog-forming emissions. This created a new market for ethanol as an oxygenate.

Oxygenated Fuels Program

The OXY Program took effect in November 1992. The program is mandated for all metropolitan statistical areas designated by the U.S. Environmental Protection Agency (EPA) as being in non-attainment for carbon monoxide. Thirty-nine metropolitan areas were originally slated to participate in the program, though less than half now remain in the program, since several have been redesignated as no longer being in non-attainment (i.e., they have met carbon monoxide emission standards). The program was implemented by the individual states in which the program areas are located. Gasoline sold in a program area must have a minimum average oxygen content of 2.7% by weight. However, by averaging, a company may sell some quantities of gasoline containing less than 2.7% so long as it sells enough gasoline containing greater than the required amount to bring the average for the total volume of fuel it sells up to 2.7%. The EPA recommended a 2.0% minimum oxygen content requirement. The oxygen requirement is in effect in each program area during its high ambient carbon monoxide period, defined to last at least four months, which span the winter months in most cities. This period is referred to as the control period. The EPA has established the specific control period for each program area.

Reformulated Gasoline Program

The RFG Program is targeted at reducing ground-level ozone pollution (i.e., smog) and lowering the levels of toxic and aromatic substances in gasoline. It took effect on January 1, 1995. Ten metropolitan areas with severe smog problems are required to participate in the program. The original areas included Los Angeles, New York, Chicago, Houston, Milwaukee, Baltimore, San Diego, Philadelphia

and Hartford. Sacramento was more recently added to the program areas. Other areas designated as being in serious, moderate or marginal ozone nonattainment were allowed to apply to opt into the program. The program is in effect year-round (i.e., the control period is 12 months). The RFG program is implemented at the federal level, rather than at the state level. Gasoline sold in a program area must have a minimum average oxygen content of 2.0% by weight.

During Phase I of the RFG program, which lasted from 1995 through 1999, volatile organic compounds (VOCs) and toxic substances had to be reduced 15% compared to conventional gasoline. Emissions of nitrogen oxides (NOx) from RFG had to be no greater than they were from conventional gasoline. The aromatic hydrocarbon content of RFG had to average no more than 25%.

Under Phase II, which took effect on January 1, 2000, the following changes were made:

- The required VOC reduction was set at 25%. (In July 2001, the EPA granted Chicago and Milwaukee an adjustment to the VOC requirement due to their reliance on ethanol as an oxygenate.)
- The required reduction of toxic pollutants was expanded to 21.5%.
- NOx emissions must be reduced by at least 1.5% (6.8% in gasoline designated as VOC-controlled).

Ethanol Incentive Programs
Federal Tax Incentives

Historically, federal and state incentives have been needed to make ethanol price-competitive in the U.S. fuel market. Currently, there are three federal "tax subsidies" available for the production and use of alcohol transportation fuels. These include: 1) a partial exemption from the federal gasoline excise taxes (the most important incentive for ethanol); 2) an income tax credit for alcohol fuels; and 3) a tax deduction for clean-fueled vehicles that use 85% alcohol (E85) fuels.

Partial Exemption from the Federal Excise Tax for Alcohol Fuels

The primary federal incentive is the exemption of 10%-ethanol blends from $0.053 of the $0.184 federal excise tax on each gallon of motor fuel.[1] Because the exemption applies to 10% blends, it amounts to an effective subsidy of $0.53 per gallon of pure ethanol ($0.053 ÷ 10%). Additionally, since January 1993, ethanol-gasoline blends consisting of 7.7% or 5.7% alcohol have received a prorated exemption. These blends, respectively, correspond to the 2.7% and 2.0% oxygen content standards for gasoline sold in OXY and RFG Program areas. The partial exemption from the tax on motor fuels applies only to alcohol derived from renewable resources such as corn, and the alcohol must be at least 190-proof (95% pure alcohol, determined without regard to any added denaturants or impurities). Ethanol and methanol qualify for this exemption without any further processing. Fuels that contain a minimum of 85% alcohol also qualify for the excise tax exemption, as does ethanol derived from other biomass sources (i.e., other than grain). The market for these fuels, however, is very small. Even though the exemption for methanol is significantly higher than that of ethanol, very little, if any, methanol is produced from renewable sources, because it is generally not cost competitive against methanol from nonrenewable sources.

Federal Income Tax Credits for Alcohol Fuels

There are three categories of income tax credits associated with ethanol: the alcohol blender's tax credit, the straight alcohol fuel credit, and the small ethanol producer's credit. The *alcohol blender's tax credit* is equivalent to the excise tax exemption and is on the same phased-in reduction schedule. Currently, alcohol blenders may receive an income tax credit of 53 cents per gallon of ethanol that they use to produce fuel. The alcohol blender's tax credit is scheduled to expire December 31, 2007. The *straight alcohol credit* applies to qualified mixtures of 85% or more alcohol. It is available only to the user of the fuel in a trade or business, or to the retail seller, as long as the fuel is used for the purpose of motor transportation. The rate for this tax credit is again 53 cents per gallon of ethanol used.

The *small ethanol producer's credit* is 10 cents per gallon of ethanol produced, used, or sold for use as a transportation fuel. This credit is limited to 15 million gallons of annual alcohol production per small producer, defined as having a production capacity of less than 30 million gallons. This credit is strictly a production tax credit available only to the manufacturer that sells alcohol to another entity for blending into a qualified mixture, for use in its own business, or for sale at retail to be used as fuel. All the income tax credits are reduced by any excise tax exemptions claimed on the same fuel. Thus, the net value of the credit is less than the nominal value, which is a key reason that the excise tax exemption is utilized much more extensively than the income tax credits.

Federal Income Tax Deduction for Alternative Fueled Vehicles

The current legislative and administration programs operate under the general umbrella of the 1992 Energy Policy Act. That legislation established a national goal of 30% penetration of U.S. light-duty vehicle fuel markets by alternative fuels, including ethanol, by 2010. It created a new federal tax deduction for individuals or businesses that purchase vehicles burning clean-fuels, including straight-alcohol fuels, and the cost of converting vehicles to operate on such fuels.

Federal Bioenergy Program

In August 1999, the Clinton Administration established the goal of tripling current domestic use of bio-based products and bioenergy by 2010. The program has the objectives of reducing emissions and adding $15–$20 billion to farm income. The Bioenergy Program was created to promote the industrial consumption of selected agricultural commodities in the production of biofuels. The USDA's Commodity Credit Corporation (CCC) provided up to $100 million in 2000 and $150 million in 2001, in incentive payments to encourage increased production of fuel grade ethanol and biodiesel. Under this program, the USDA will make up to $150 million in payments to commercial ethanol and biodiesel producers that increase their bioenergy production between October 1, 2001, and September 30, 2002. Payments will be based on increases in the production of bioenergy from eligible commodities[2] compared to the same time period a year earlier.

Federal Biomass Energy Programs

Cellulosic ethanol can be produced from a range of biomass feedstocks, including rice straw, agricultural residues, and dedicated energy crops like switch grass and fast-growing trees. The USDA and DOE conduct research on ethanol processes and biodiesel development. The legislation encourages the evaluation of new energy crops and accelerates the development of advanced biomass technologies to produce a variety of energy-related products and reduce U.S. reliance on fossil fuels.

Trade and Development Act of 2000

This law is aimed at expanding two-way trade and creating incentives for the countries of the Caribbean Basin to continue reforming their economies. This legislation removed the rules of origin requirements for ethanol imported into the U.S. from Caribbean Basin Initiative (CBI) countries.

Agriculture, Rural Development, Food and Drug Administration and Related Agencies Appropriations Act of 1999

Section 769 of this Act authorizes the Secretary of Agriculture to approve not more than six projects (no more than one in any state) on land subject to Conservation Reserve Program (CRP) contracts in which crops may be harvested for biomass used in energy production.

Regional Biomass Energy Program

The initiative required the DOE to support regional biomass energy programs and identify a plan for regionally appropriate biomass technologies. The RBEP budget is modest, at about $4 million annually since 1983, but it helps support nearly $12 million in projects annually.

Proposed Federal Ethanol Programs

Securing America's Energy Future Act of 2001

The bill is intended to support the security and diversity of the U.S. energy supply. It includes a section entitled the Bioenergy Act of 2001 and authorizes $150 million for biomass-related activities in 2002 and increases funding to $220 million by 2006.

DOE Fiscal Year 2002 Biofuels Energy Systems Program

The Department of Energy has requested $81.955 million for its biomass and biopower energy systems budget for FY 2002. The Biofuels Energy Systems program has identified ethanol as the most promising of the liquid transportation fuels options in the near and mid-term.

USDA Fiscal Year 2002 New Uses for Agricultural Products

In the proposed USDA budget for fiscal year 2002, the Bush Administration supports investments in new technologies to develop advanced products based on agricultural commodities, for markets in the U.S. and abroad.

State Incentives

Some states have a variety of incentives for the production and use of fuel ethanol, ranging from excise tax exemptions to producer payments. Incentives for states in the region are detailed in Table 1.

Ongoing Policy Issues
Reformulated Gasoline and MTBE

MTBE has been the primary oxygenate used in the RFG Program. However, MTBE also has contaminated groundwater in a number of locations around the country.

Table 1 Ethanol Incentives by State

	Ethanol Incentive
Iowa	1 ¢/gal - excise exemption
Minnesota	20 ¢/gal - producer payment
Missouri	20 ¢/gal - producer payment
Montana	30 ¢/gal - producer payment
Nebraska	20 ¢/gal - producer payment
South Dakota	20 ¢/gal - producer payment
Wyoming	40 ¢/gal - producer payment

Source: Oxy-fuel News, August 30, 2001.

Accordingly, MTBE is being viewed by some policymakers as causing a tradeoff between the goals of clean air and clean water. The EPA does not regulate the levels of MTBE in drinking water; however, large public water utilities are required to monitor for the compound, and some states have set individual standards for acceptable MTBE levels in public drinking water.

The slow pace that science has taken to prove the adverse affects of MTBE has slowed the debate over legislation banning the use of the additive nationally. Two legislative bills are now pending which would eliminate MTBE as a fuel additive. It is likely that one of these pending bills will be passed by the end of the 107th Congress, putting the timetable on an all-out federal ban on MTBE in the United States between three and four years.

The result of the lack of clear federal action has been a stream of announcements by state governments that they would phase out and/or ban the use of MTBE, as well as proposals by Congress and the EPA to update the CAA90, the Clean Water Act, and associated regulations.

Phase 2 Reformulated Gasoline

As part of the Phase 2 requirements of the RFG Program, which took effect in 2000, gasoline sold in the summer months (beginning June 1) must meet tighter volatility standards. Because of its physical properties, ethanol has a higher RVP than MTBE. Thus, Phase 2 RFG-compliant fuels using ethanol require base gasoline with a lower RVP, referred to as Reformulated Gasoline Blendstock for Oxygenate Blending (RBOB), which is more expensive to produce. To reduce the cost of ethanol-blended RFG and decrease the potential for price spikes, the EPA issued a rulemaking in June 2001 adjusting the VOC standards for RFG in Chicago and Milwaukee, the only areas of the country that use ethanol exclusively in RFG. This VOC adjustment is equivalent to a slightly higher RVP allowance.

New legislative options beyond Phase 2 have included eliminating the oxygenate standard for RFG, or suspending the program entirely. However, some in the petroleum industry suggest that additional changes to fuel requirements could further disrupt gasoline supplies. Four current bills would allow a higher RVP for ethanol-blended fuels. These are H.R. 454 (Johnson, T.), H.R. 1999 (Nussle), S. 670 (Daschle), and S. 892 (Harkin). As of the writing of this report, all four have been referred to committee, but no hearings or markups have been held.

Ethanol Tax Incentives

The incentives that allow fuel ethanol to compete with other additives continue to be controversial. Nevertheless,

Congress in 1998 extended the motor fuels tax exemption through 2007, but at declining rates.

In the current Congress, S. 907 (Carnahan) would extend the alcohol fuels tax exemption through 2015. In addition, several bills would expand the availability of the small producer credit, increase the size of a covered producer, and make the credit available to cooperatives. These include H.R. 1636 (Thune), H.R. 1999 (Nussle), S. 312 (Grassley), S. 613 (Fitzgerald), and S. 907 (Carnahan). All have been referred to committee, but no markups have been held. A hearing was held on S. 312.

H.R. 2303 (Lewis, Ron) contains the above provisions on small producers and cooperatives. In addition, the bill would provide tax credits for the retail sale of ethanol and for the installation of retail infrastructure. This bill has been referred to committee, but no hearings or markups have been held.

Renewable Fuels Standard

A renewable fuels standard (RFS) would require that the nation's fuel supply contain a certain percentage of renewable fuels. The definition of a renewable fuel would include biodiesel, ethanol, or any other liquid fuel produced from biomass or biogas that is used to reduce the quantity of fossil fuel present in a fuel mixture. The RFS would be measured in gasoline-equivalent gallons and would cover both gasoline and diesel fuel.

Under the Hagel/Johnson legislation the renewable fuels standard that is being purposed. In this bill currently is very aggressive. Under this bill, the renewable fuels standard would start off at a low .80% of total U.S. gasoline consumption in 2002 and rise every year until it reaches a level of 5% by 2016. In the Daschle/Lugar Bill, the renewable fuels requirement starts at 0.60% in 2002 and increases up to a maximum rate of 1.5% in 2011.

Farm Policy Considerations

United States policy provides support to the agriculture sector in a wide variety of ways, including direct payments to support producer incomes, beneficial loan programs, investment in infrastructure, and support for soil conservation and water quality. The bulk of government program support for agriculture goes for a few crops. For example, USDA outlays for direct commodity support in FY 2000 amounted to $16.1 billion, of which corn accounted for $5.77 billion. Corn, cotton, rice, soybeans and wheat together accounted for $14.4 billion, just under 90% of the total. The future direction and size of federal agriculture programs are extremely uncertain as of the writing of this report.

U. S. Ethanol Production
Background

As was depicted in the chart contained in the introduction to this report (Figure 1), the production of ethanol in the United States has expanded greatly since the late 1970s. Ethanol production stood at 175 million gallons in 1980, grew to 900 million gallons by 1990, and reached a record 1.63 billion gallons in 2000. Thus, ethanol production has expanded at a rate of 725–730 million gallons per decade over each of the last two decades, equivalent to a compounded annual growth rate of 11.2%. Excluding the rebound in 1997 from the plant shutdowns that were induced by high corn prices and tight availability in 1996, the highest year-over-year increases in ethanol production have been approximately 150–160 million gallons. (Current ethanol production for the year 2002 is forecast at about 2.3 billion gallons, an increase of over 535 million gallons from the prior year.)

Current Industry Structure

The ethanol industry is geographically concentrated in the Corn Belt (see Figure 1). Specifically, most plants—and a large majority of capacity—are located in a triangle that stretches from south central Nebraska to central Minnesota over to eastern Illinois (see Figure 2). Significant plant additions in Nebraska and Minnesota have occurred during the last 10–15 years due to corn supplies and state incentives.

The current capacity of the ethanol industry is estimated to be almost exactly 2.5 billion gallons. Archer Daniels Midland (ADM) remains the largest ethanol producer, able to produce 950 million gallons per year (mgy), or 40% of total U.S. capacity (see Table 1). ADM's share of U.S. capacity has decreased moderately over the last decade, as both large corporations and farmer-owned cooperatives have entered the industry at a pace faster than ADM's expansion. ADM has chosen to grow predominantly through the expansion of its existing mammoth facilities rather than through the construction of new ones, due to (1) ADM's reliance on wet mills, (2) the economies of scale possible for wet mills, and (3) the lower cost per gallon of expanding a facility versus constructing a new one. ADM is roughly seven times as large as the next-largest companies in the industry (Cargill, Minnesota Corn Processors and Williams Bio-Energy), each of which has 100–112 mgy of capacity.

Future Capacity

According to the Renewable Fuels Association, seventeen ethanol plants are currently under construction.

112

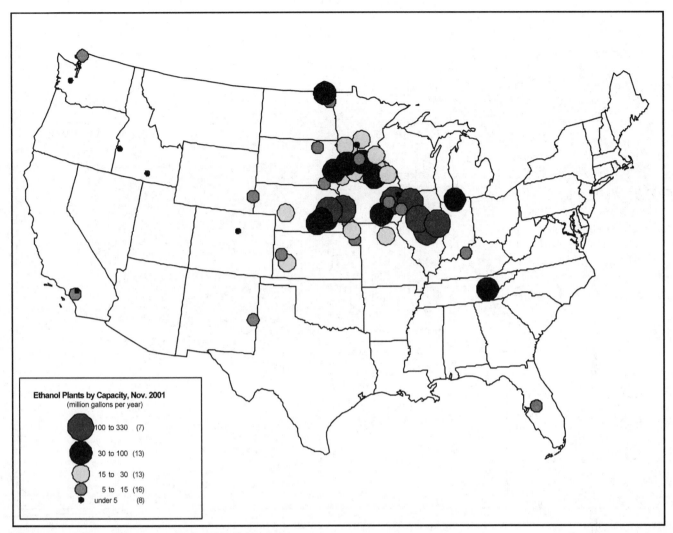

Ethanol Plants by Capacity, Nov. 2001
(million gallons per year)

100 to 330 (7)

30 to 100 (13)

15 to 30 (13)

5 to 15 (16)

under 5 (8)

Figure 2 Ethanol plants by capacity, November 2001.

Almost all are new entrants to the industry, and many are owned by farmer organizations. All are located in or near the geographic "triangle" where the existing industry is concentrated (see Figure 3). The capacities of the individual facilities range from 0.7 to 45 mgy, and the total capacity is 337 mgy (see Table 2). This would bring U.S. capacity to 2.5 billion gallons per year (bgy). In addition to these plant there are proposed facilities under discussion or development in virtually every part of the country.

Great Plains Ethanol Situation and Outlook

With the current and planned buildup in production capacity and a significant location advantage, the Great Plains region stands to enjoy strong economic gains due to the expansion of the ethanol industry in the region. The move is particularly timely as it comes amidst the legislative push to eliminate fuel additives that are sought unsafe in many additional areas of the U.S.

Figure 5 shows the planned expansion in ethanol production capacity to come on stream in the next few years. Five plants, each in the 40 to 45 million gallon capacity range, are planned for Iowa and South Dakota. Three additional plants with 15 to 20 million gallon capacity are planned in Iowa and Nebraska.

Figure 6 depicts the volume of corn production and planned ethanol plants by capacity of production. The darker shaded areas reveal counties where corn production is concentrated. Gray circles show ethanol plants that were operational as of November 2001 while white circles identify those plants intended to come into production over the next two to three years. Thus in a fairly short time span production capacity could be expanded by 300 to 500 million gallons annually with these new plants. The volume of corn production in the region does not indicate any potential problems with raw product supply for ethanol production.

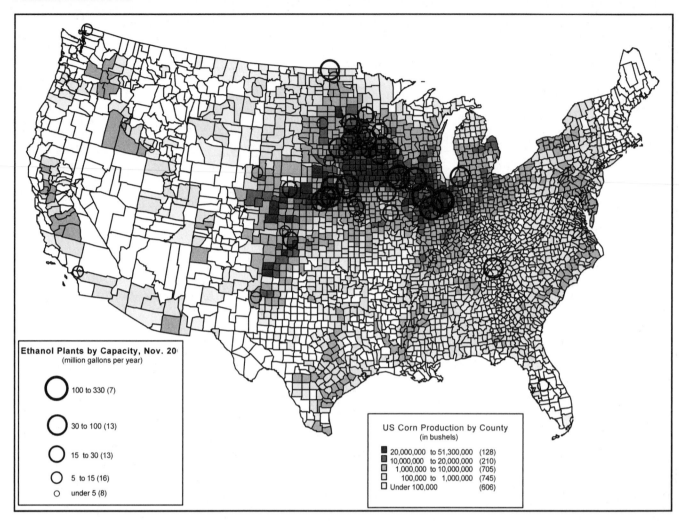

Figure 3 Relationship of ethanol processing facilities to major corn producing regions
Note: all types of processing facilities (e.g., whey, wood pulp) are included on the map.

Table 2   Operational Ethanol Facilities

Company Name	City	State	Feedstock	Capacity (mgy)
A.E. Staley	Loudon	TN	Corn	60.0
AGP	Hastings	NE	Corn	52.0
Agri-Energy	Luverne	MN	Corn	17.0
Alchem	Grafton	ND	Corn	10.5
Al-Corn Clean Fuel	Claremont	MN	Corn	17.0
Archer Daniels Midland*	Decatur	IL	Corn	330.0
Archer Daniels Midland*	Peoria	IL	Corn	110.0
Archer Daniels Midland*	Cedar Rapids	IA	Corn	212.0
Archer Daniels Midland*	Clinton	IA	Corn	115.0
Archer Daniels Midland*	Walhalla	ND	Corn/Barley	30.0
Broin Enterprises	Scotland	SD	Corn	7.0
Cargill	Blair	NE	Corn	75.0
Cargill	Eddyville	IA	Corn	35.0
Central Minnesota	Little Falls	MN	Corn	18.0
Chief Ethanol	Hastings	NE	Corn	62.0

(continued)

Table 2 Operational Ethanol Facilities (continued)

Company Name	City	State	Feedstock	Capacity (mgy)
Chippewa Valley Ethanol	Benson	MN	Corn	20.0
Corn Plus	Winnebago	MN	Corn	20.0
Dakota Ethanol	Wentworth	SD	Corn	40.0
DENCO, LLC.	Morris	MN	Corn	17.0
ESE Alcohol	Leoti	KS	Seed Corn	1.5
Ethanol 2000	Bingham Lake	MN	Corn	28.0
Exol, Inc.	Albert Lea	MN	Corn	18.0
Georgia-Pacific	Bellingham	WA	Paper Waste	7.0
Golden Cheese	Corona	CA	Whey	5.0
Golden Triangle Energy	Craig	MO	Corn	15.0
Gopher State Ethanol	St. Paul	MN	Corn	15.0
Grain Processing Corp.	Muscatine	IA	Corn	10.0
Heartland Corn Products	Winthrop	MN	Corn	35.0
Heartland Grain Fuel	Aberdeen	SD	Corn	8.0
Heartland Grain Fuel	Huron	SD	Corn	14.0
High Plains Corporation	York	NE	Corn/Milo	50.0
High Plains Corporation	Colwich	KS	Corn	20.0
High Plains Corporation	Portales	NM	Corn	14.0
J.R. Simplot	Caldwell	ID	Potato Waste	3.0
J.R. Simplot	Burley	ID	Potato Waste	3.0
Kraft, Inc.	Melrose	MN	Whey	2.6
Manildra Ethanol	Hamburg	IA	Corn/Milo/Wheat Starch	7.0
Merrick/Coors	Golden	CO	Waste Beer	1.5
Midwest Grain	Pekin	IL	Corn/Wheat Starch	68.0
Midwest Grain	Atchison	KS	Corn	10.0
Miller Brewing	Olympia	WA	Brewery Waste	0.7
Minnesota Corn Proc.	Columbus	NE	Corn	90.0
Minnesota Corn Proc.	Marshall	MN	Corn	32.0
Minnesota Energy	Buffalo Lake	MN	Corn	12.0
Nebraska Energy	Aurora	NE	Corn	30.0
New Energy Co.	South Bend	IN	Corn	85.0
Northeast MO Grain Proc.	Macon	MO	Corn	15.0
Parallel Products	Louisville	KY	Beverage Waste	7.0
Parallel Products	Bartow	FL	Beverage Waste	5.0
Parallel Products	R. Cucamonga	CA	Beverage Waste	3.0
Permeate Refining	Hopkinton	IA	Sugars & Starches	1.5
Pro-Corn	Preston	MN	Corn	22.0
Reeve Agri-Energy	Garden City	KS	Corn/Milo	10.0
Sunrise Energy	Olympia	IA	Corn	7.0
Sutherland Associates	Sutherland	NE	Corn	15.0
Williams Bio-Energy Svcs.	Pekin	IL	Corn	100.0
Total				2,018.3

Sources: BBI Int'l, Renewable Fuels Assn., Milling & Baking News, Industry Sources
*ADM recently increased total capacity to 797 mgy. Plant sizes were known when capacity was 750 mgy and prorated for the increase.

Table 3 denotes the company, city and state where ethanol plants are under construction and their potential capacity. The bulk of the new plants intend to utilize corn as the feedstock. The list of proposed plants changes frequently as construction is completed. Currently the Renewable Fuels Association lists 8 plants that are on the books for Iowa, South Dakota and Nebraska. When these plants are completed, a total capacity increase of 255 million gallons annually would be anticipated.

Table 4 shows the forecast for U.S. ethanol production capacity through 2005 with projections for both expansions of capacity by existing companies and new entrants

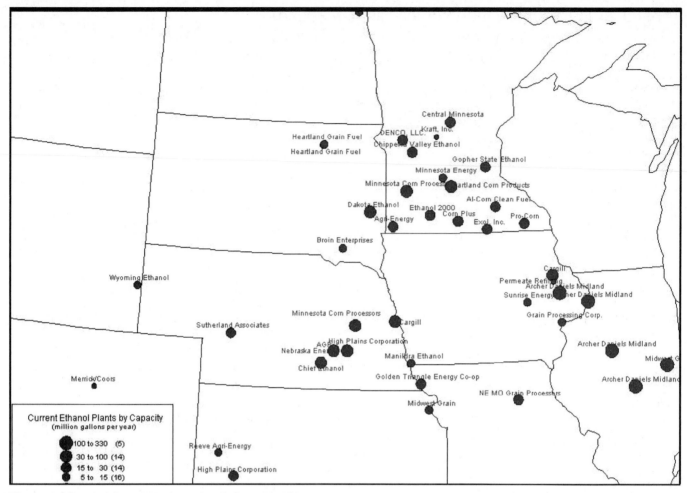

Figure 4 Great plains region operational ethanol facilities

Table 3 Ethanol Plants Under Construction

Company Name	City	State	Feedstock	Capacity
Glacial Lakes Energy, LLC	Watertown	SD	Corn	40
Husker Ag Processing	Plainview	NE	Corn	20
Little Sioux Corn Processors, LLC	Marcus	IA	Corn	40
Midwest Grain Processors	Lakota	IA	Corn	45
Northeast Iowa Ethanol, LLC	Earlville	IA	Corn	15
Northern Lights Ethanol, LLC	Milbank	SD	Corn	40
Pine Lake Corn Processors, LLC	Steamboat Rock	IA	Corn	15
Tall Corn Ethanol, LLC	Coon Rapids	IA	Corn	40
Total				255

Sources: BBI International, Renewable Fuels Assn.

to the industry. The bulk of the expected growth (70%) is expected to come from new entrants as the number of plants is forecast to increase by 46 plants during the period. Current low corn prices and moderate oil prices have supported the expansion of capacity, as has anticipation that California will consume significant quantities of ethanol within the next couple of years.

Table 4 shows existing and future capacity specifically in the Great Plains region and the surrounding Corn Belt states. As one might anticipate with corn as the major feedstock, those states with large corn production are leaders in ethanol production and planned expansion. Iowa has the most significant expansion plans with respect to volume with South Dakota a distant second.

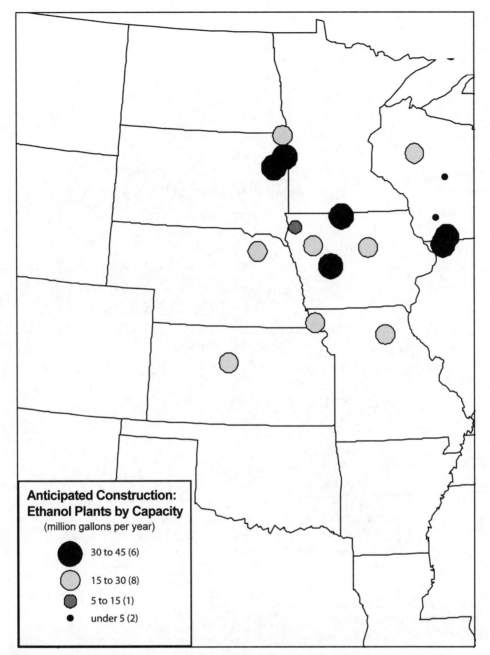

Figure 5 Anticipated new construction: ethanol plants by capacity

Table 4 Planned Growth of Industry Production Capacity (Cumulative by End of Year)

		2001	2002	2003	2004	2005
Existing Industry	Companies	44	44	44	44	44
	Plants	57	58	58	58	58
	Capacity (mgy)	2,219	2,481	2,689	2,774	2,852
New Entrants	Companies	4	21	40	40	40
	Plants	4	21	43	44	46
	Capacity (mgy)	82	518	1,329	1,387	1,575
Existing & New Plants	Companies	48	65	84	84	84
	Plants	61	79	101	102	104
	Capacity (mgy)	2,301	2,999	4,018	4,161	4,427

Source: California Energy Commission.

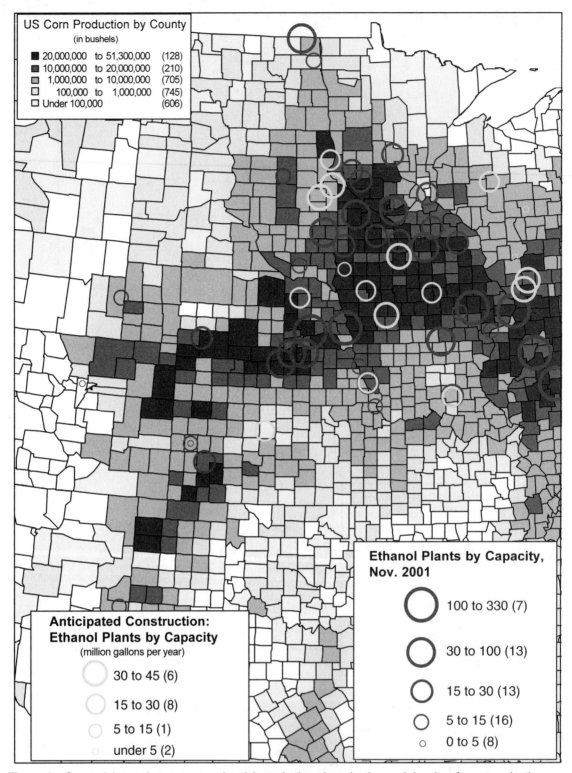

Figure 6 Great plains region: current and anticipated ethanol production and density of corn production.

Kansas and Nebraska each intend to expand capacity by significant percentages, but the level of volume is still small relative to Iowa. It is interesting to note that no expansion in existing capacity is planned in several other states in the region with large corn production, such as Nebraska and Minnesota.

This reflects the entry of 40 new companies into the ethanol industry, in addition to the 44 currently operating. If realized, the industry would double in terms of both the number of participants and their aggregate capacity. However, the Commission issued the caveat: "While survey respondents were asked to provide data on fuel-grade

Table 5   Anticipated Increase in Ethanol Processing Capacity
in the Great Plains, Million Gallons per Year (mmgy)

Region	Existing Capacity mmgy	Capacity Under Construction mmgy	% Change in Capacity	Future Capacity mmgy
Colorado	1.5	0.0	0.0%	1.5
Iowa	422.5	132.0	31.2%	554.5
Kansas	41.5	25.0	60.2%	66.5
Oklahoma	0.0	0.0	0.0%	0.0
Minnesota	291.6	0.0	0.0%	291.6
Missouri	30.0	0.0	0.0%	30.0
North Dakota	40.5	0.0	0.0%	40.5
Nebraska	453.0	20.0	4.4%	473.0
South Dakota	69.0	95.0	137.7%	164.0
Great Plains	1,349.6	177.0	13.1%	1,526.6
Rest of the Country	800.7	99.7	12.5%	900.4
US	2,150.3	276.7	12.9%	2,427.0

ethanol production capacity only, some amount (presumably not substantial) of ethanol production for the industrial or beverage markets may have been included in the tabulated survey results." It is also unknown whether respondents included the portion of the grind that would normally be dedicated to production of corn sweeteners at facilities with so-called swing capacity (i.e., that typically produce high fructose corn syrup during warmer months and ethanol during cooler months).

Notes

1. 26 U.S.C. 40.

2. Eligible commodities for FY 2002 are barley, corn, grain sorghum, oats, rice, wheat, soybeans, sunflower seed, canola, crambe, rapeseed, safflower, sesame seed, flaxseed, mustard seed, and cellulosic crops (such as switchgrass and short rotation trees) grown on farms in the United States and its territories.

Clogged Arteries

America's aging and congested road, rail, and air networks are threatening its economic health.

Bruce Katz and Robert Puentes

Transportation spending is spread around the United States like peanut butter, and while it's spread pretty thick—nearly $50 billion last year in federal dollars for surface transportation alone—the places that are most critical to the country's economic competitiveness don't get what they need. The nation's 100 largest metropolitan regions generate 75 percent of its economic output. They also handle 75 percent of its foreign sea cargo, 79 percent of its air cargo, and 92 percent of its air-passenger traffic. Yet of the 6,373 earmarked projects that dominate the current federal transportation law, only half are targeted at these metro areas.

In the past, strategic investments in the nation's connective tissue—to develop railroads in the 19th century and the highway system in the 20th—turbocharged growth and transformed the country. But more recently, America's transportation infrastructure has not kept pace with the growth and evolution of the economy. As earmarks have proliferated, the government's infrastructure investment has lost focus. A recent academic study shows that public investment in transportation in the 1970s generated a return approaching 20 percent, mostly in the form of higher productivity. Investments in the 1980s generated only a 5 percent return; in the 1990s, the return was just 1 percent.

The map above shows an estimate of road-traffic congestion in 2010. In most major metro areas, it is steadily worsening. The cost of congestion, including added freight cost and lost productivity for commuters, reached $78 billion in 2005. Half of that occurred in just 10 metro areas.

America's biggest and most productive metro regions gather and strengthen the assets that drive the country's prosperity—innovative firms, highly productive and creative workers, institutions of advanced research. And the attributes of some cities are not easily replicated elsewhere in the U.S. The most highly skilled financial professionals, for instance, do not choose between New York and Phoenix. They choose between New York and London—or Shanghai. While many factors affect that choice, over time, the accretion of delays and travel hassles can sap cities of their vigor and appeal. Arriving at Shanghai's modern Pudong airport, you can hop aboard a maglev train that gets you downtown in eight minutes, at speeds approaching 300 miles an hour. When you land at JFK, on the other hand, you'll have to take a train to Queens, walk over an indoor bridge, and then transfer to the antiquated Long Island Rail Road; from there, downtown Manhattan is another 35 minutes away.

To power our metropolitan engines, we need to make big, well-targeted investments that improve transportation within and around them. Above all else, that means taking a less egalitarian approach to our infrastructure: there is little justification for making small improvements all over the place.

In a post-agricultural, postindustrial, innovation-dependent economy, the roads to prosperity inevitably pass through a few essential cities. We should make sure they're well maintained.

Bruce Katz is the director of the Brooking Institution's Metropolitan Policy Program. **Robert Puentes** is a fellow in the Metropolitan Policy Program.

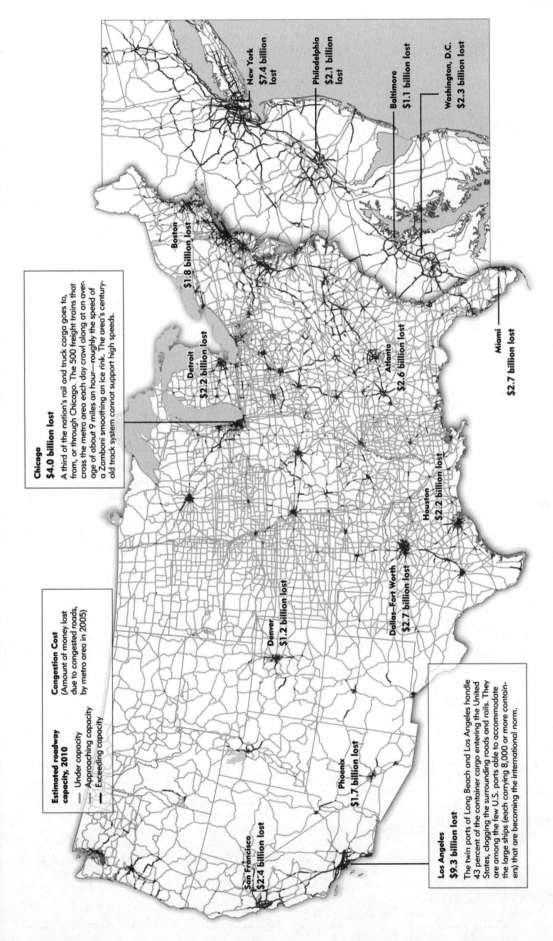

Estimated roadway capacity, 2010
— Under capacity
— Approaching capacity
— Exceeding capacity

Congestion Cost
(Amount of money lost due to congested roads, by metro area in 2005)

Chicago
$4.0 billion lost

A third of the nation's rail and truck cargo goes to, from, or through Chicago. The 500 freight trains that cross the metro area each day crawl along at an average of about 9 miles an hour—roughly the speed of a Zamboni smoothing an ice rink. The area's century-old track system cannot support high speeds.

Los Angeles
$9.3 billion lost

The twin ports of Long Beach and Los Angeles handle 43 percent of the container cargo entering the United States, clogging the surrounding roads and rails. They are among the few U.S. ports able to accommodate the large ships (each carrying 8,000 or more containers) that are becoming the international norm.

New York
$7.4 billion lost

Philadelphia
$2.1 billion lost

Baltimore
$1.1 billion lost

Washington, D.C.
$2.3 billion lost

Boston
$1.8 billion lost

Detroit
$2.2 billion lost

Atlanta
$2.6 billion lost

Miami
$2.7 billion lost

Houston
$2.2 billion lost

Denver
$1.2 billion lost

Dallas–Fort Worth
$2.7 billion lost

Phoenix
$1.7 billion lost

San Francisco
$2.4 billion lost

Manifest Destinations

MARC BAIN AND KEVIN HAND

We call this country the land of opportunity, and millions of us never give up the search for it. Where richer land, milder weather and open spaces lured the first colonists west and south, the same overall migration pattern continues, largely driven now by dreams of better places to work and play. The fastest population growth tends to be in the suburbs.

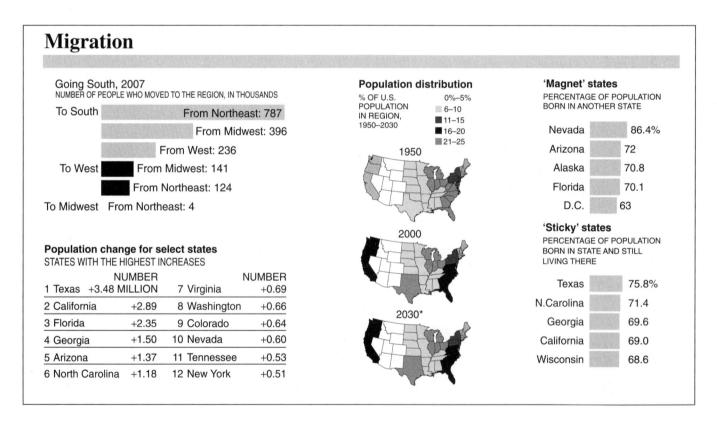

Migration

Going South, 2007
NUMBER OF PEOPLE WHO MOVED TO THE REGION, IN THOUSANDS

To South — From Northeast: 787
From Midwest: 396
From West: 236

To West — From Midwest: 141
From Northeast: 124

To Midwest — From Northeast: 4

Population change for select states
STATES WITH THE HIGHEST INCREASES

	NUMBER			NUMBER
1 Texas	+3.48 MILLION	7	Virginia	+0.69
2 California	+2.89	8	Washington	+0.66
3 Florida	+2.35	9	Colorado	+0.64
4 Georgia	+1.50	10	Nevada	+0.60
5 Arizona	+1.37	11	Tennessee	+0.53
6 North Carolina	+1.18	12	New York	+0.51

Population distribution
% OF U.S. POPULATION IN REGION, 1950–2030

0%–5%
6–10
11–15
16–20
21–25

1950

2000

2030*

'Magnet' states
PERCENTAGE OF POPULATION BORN IN ANOTHER STATE

Nevada	86.4%
Arizona	72
Alaska	70.8
Florida	70.1
D.C.	63

'Sticky' states
PERCENTAGE OF POPULATION BORN IN STATE AND STILL LIVING THERE

Texas	75.8%
N.Carolina	71.4
Georgia	69.6
California	69.0
Wisconsin	68.6

Urban Sprawl

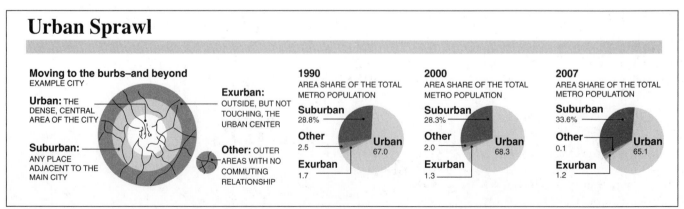

Moving to the burbs–and beyond
EXAMPLE CITY

Urban: THE DENSE, CENTRAL AREA OF THE CITY

Suburban: ANY PLACE ADJACENT TO THE MAIN CITY

Exurban: OUTSIDE, BUT NOT TOUCHING, THE URBAN CENTER

Other: OUTER AREAS WITH NO COMMUTING RELATIONSHIP

1990
AREA SHARE OF THE TOTAL METRO POPULATION
Suburban 28.8%
Other 2.5
Urban 67.0
Exurban 1.7

2000
AREA SHARE OF THE TOTAL METRO POPULATION
Suburban 28.3%
Other 2.0
Urban 68.3
Exurban 1.3

2007
AREA SHARE OF THE TOTAL METRO POPULATION
Suburban 33.6%
Other 0.1
Urban 65.1
Exurban 1.2

Major Metros

The urban share of the American economy

METRO AREAS ACCOUNT FOR ONLY 12 PERCENT OF THE LAND AREA IN THE U.S., YET THE PRODUCE 75 PERCENT OF THE GDP, THEIR SHARE OF SELECT ECONOMIC SECTORS:

Economy sectors

- ■ POPULATION AND ECONOMY
- ■ HUMAN CAPITAL
- ■ INNOVATION
- ■ INFRASTRUCTURE

Land area: 12%

Population: 65

Research universities: 67

Jobs: 68

Graduate-degree holders: 75

Knowledge-economy jobs: 76

Patents: 78

Air cargo: 79

R&D employment: 81

Air-passenger boardings: 92

Venture capital funding: 94

Public-transit passenger miles: 95

Major metro areas with the largest share of select Industries

● Manufacturing capitals
SHARE OF INDUSTRY'S TOTAL PAYROLL[†]

		% SHARE
❶	Biomedical	21.8
❷	Technology	20.7
❸	Pharmaceuticals	20.1
❹	Motor-vehicle parts	18.0
❺	Aerospace	16.5

IPAY INCLUDES WAGES, VACATION, BONUSES, STOCK OPTIONS AND TIPS.

Seattle

Minneapolis/St. Paul

San Jose

Los Angeles

Houston

Detroit

New York, New Jersey, Long Island

Washington, D.C.

● Service capitals
SHARE OF INDUSTRY'S TOTAL PAYROLL[†]

		% SHARE
❶	Finance	42.7
❷	Legal services	14.7
❸	Energy	12.1
❹	Consulting	8.7
❺	IT	6.9

AIDS Infects Education Systems in Africa

Yet school is critical factor in combating pandemic.

BESS KELLER

Teachers in Zambia nearly went on strike a few years ago because they weren't paid on time.

Neither of the two government officials responsible for the payroll had reported to work as the salaries came due. One was out sick, almost certainly from an AIDS-related illness. The other was attending to the death rites of someone who had died of the disease.

A strike was averted only when the funeral had taken place and the second official returned to work.

The Rev. Michael J. Kelly, a longtime professor of education at the University of Zambia until his recent retirement, tells this story to bring home the point that the AIDS pandemic raging across sub-Saharan Africa doesn't stop with personal carnage. It also threatens whole systems, including what is arguably the most critical for the region's future—education.

"For countries that are not very rich in managerial personnel, the loss of a few managers can be very deleterious," Father Kelly said.

Similarly, the loss of teachers in the region can sink the quality of education as classes are combined and teachers with fewer qualifications are hired.

"In many African countries, we are talking about educational systems that are already fragile in many ways—they may lack materials; teachers are poorly trained, especially in rural areas," said Cream Wright, the chief of education for UNICEF and a native of Sierra Leone.

In such a situation, any reversal—the death of a teacher, the bereavement of a child, a reduction in budget—can more easily do harm.

The injury goes deeper than it otherwise might because AIDS is destroying families, which undergird the education system. Families are the mainstay of schooling in any country, but in African nations, the family is often the only social safety net that can keep children in school. Now, even that net is seriously frayed by the AIDS-related illnesses and deaths of men and women in their most productive working years.

"Both on the side of schools and of homes, AIDS destroys those coping mechanisms that are in place," Mr. Wright said.

Where rates of HIV infection are high, as they are in much of southern and eastern Africa, experts warn, the effects on social stability and education are so great that young people are being robbed of hope, and national development is being stunted.

And in a final merciless twist, declines in education reduce the chances of arresting the pandemic, since schools may be the best way to reach uninfected young people with the information, skills, and attitudes that ultimately protect them.

11 Million and Counting

Of the estimated 39 million people worldwide living with the human immunodeficiency virus, for which there is no vaccine and no cure, some 70 percent are in sub-Saharan Africa.

Yet the prevalence of the virus that causes AIDS varies enormously even in this hardest-hit region. In several West African countries, including the most populous, Nigeria, the infection rate is less than 5 percent. But in Botswana and Swaziland in southern Africa, more than 35 percent of the adult population is infected, and the rate continues to rise, according to UNAIDS, the United Nations AIDS coordinating group.

Millions of Africans have died of the disease in the past 20 years. The bereft include 11 million sub-Saharan children who have lost one or both parents to the disease, making the total number of orphans in the region more than 34 million. The number of AIDS orphans is expected to rise to 20 million by the end of the decade.

The loss of parents affects school enrollment and learning. Families with fewer workers are less likely to be able to afford the costs of school or be able to forgo the labor of a child who is enrolled. Sick relatives make further demands on children, especially girls, who in many African countries devote hours a day to household tasks.

Juliet Chilengi knows firsthand that orphaned children are likely to lose their chance for an education. To save some of them from that fate, Ms. Chilengi founded the New Horizon orphanage in Lusaka, Zambia, where she lives. One teenage girl

Orphaned by AIDS

Millions of children have been orphaned as a result of the AIDS pandemic in sub-Saharan Africa, and the numbers are expected to climb.

	2001	2010[*]
Angola	104,000	331,000
Benin	34,000	113,000
Botswana	69,000	120,000
Burkina Faso	268,000	415,000
Burundi	237,000	296,000
Cameroon	210,000	677,000
Central African Republic	107,000	165,000
Chad	72,000	132,000
Congo	78,000	112,000
Côte d'Ivoire	420,000	539,000
Djibouti	6,000	15,000
Dem. Rep. of Congo	927,000	1,366,000
Equatorial Gunea	<100	1,000
Eritrea	24,000	55,000
Ethiopia	989,000	2,165,000
Gabon	9,000	14,000
Gambia	5,000	8,000
Ghana	204,000	263,000
Guinea	29,000	57,000
Guinea-Bissau	4,000	13,000
Kenya	892,000	1,541,000
Lesotho	73,000	169,000
Liberia	39,000	121,000
Madagascar	6,000	17,000
Malawi	468,000	741,000
Mali	70,000	117,000
Mozambique	418,000	1,064,000
Namibia	47,000	118,000
Niger	33,000	123,000
Nigeria	995,000	2,638,000
Rwanda	264,000	356,000
Senegal	15,000	23,000
Sierra Leone	42,000	121,000
South Africa	662,000	1,700,000
Sudan	62,000	373,000
Swaziland	35,000	71,000
Tanzania	815,000	1,167,000
Togo	63,000	127,000
Uganda	884,000	605,000
Zambia	572,000	836,000
Zimbabwe	782,000	1,191,000

*Projected SOURCES: UNAIDS and UNICEF

now with Ms. Chilengi arrived after the aunt with whom she lived judged the girl a handful and "chased her away." Another succumbed to a sexual liaison with an older man because she had no one else to pay for her education, which was cut short anyway when she got pregnant.

Many extended families in her country are breaking under the strain of poverty from unemployment and AIDS, Ms. Chilengi explained.

"There is such a drastic change," agreed Lucy Barimbui, who coordinates anti-AIDS activities for the Kenya National Union of Teachers. "It's no longer the Africa where a child belongs to everyone, and the teachers have to deal with that."

Even where children's material needs are met, grief and insecurity all too easily interfere with learning.

"In many African countries, we are talking about educational systems that are already fragile in many ways."

Cream Wright
Chief of Education, UNICEF

"We try to talk to them and encourage them, so they don't feel the absence of the parents," said Bartholomew Njogu, the head of Nkubu Primary School in Nkubu, Kenya. But that is a tall order for a school that has no paid counselors and where classes number around 40 children.

Taking Action

Meanwhile, the pandemic has sickened and killed thousands of trained school employees. At one point, experts identified male teachers as being a particularly at-risk population because many are posted to jobs away from their families and have the money for extramarital sexual partners—including readily available female students. In the absence of convincing evidence, however, many experts now believe that teachers are no more likely than others of their area and background to be infected.

Still, classes already swollen by recent guarantees of free primary education in, for instance, Kenya, Malawi, and Zambia, are doubled up when teachers are absent. Rural schools, which are harder to staff, particularly suffer.

To try to preserve its teaching force, Zambia has recently begun offering free antiviral therapy to infected teachers, and a pilot project of several South African teachers' unions will do the same.

Education bureaucracies are generally ill equipped to project manpower needs, and in many countries, internationally imposed fiscal constraints, as well as internal economic ones, have kept hiring of additional school staff members to a minimum. A decline in school quality going back to the 1970s that had begun to be arrested in the 1990s has once again taken hold, many experts say.

In fact, the creeping nature of the crisis erodes improvements just when education needs to be making quantum leaps. Schools should be humane and supportive environments, for instance, free from sexual harassment of girls, who far outpace boys in their rate of HIV infection. Gambia has taken a step in

Widespread Pestilence

The rates of people ages 15 to 49 living with HIV or AIDS in 2001 were highest in the southern part of the African continent.

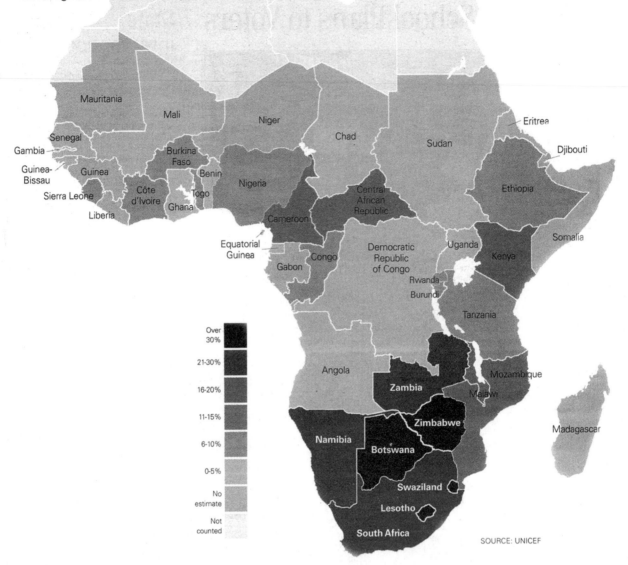

SOURCE: UNICEF

that direction by passing a national law prohibiting male teachers from allowing female students as visitors in their homes. Schools in Zambia, for their part, have been fostering compassion for people with AIDS by distributing AIDS emblems to be worn by all teachers on a certain day of the month. And Kenya has produced a pocket-size AIDS-in-education policy that is to go to all teachers.

On new fronts, some education thinkers are proposing a massive resurrection of boarding schools, which had largely fallen into disfavor with aid donors as expensive artifacts of the colonial era. Though day schools are cheaper, boarding schools might provide orphans with a nurturing community as well as

an education. Policy experts also envision more flexible forms of schooling tailored to the needs of youngsters who are caring for family members, bringing in income, or themselves heading families.

"This generation is being forced to take on adult responsibilities earlier than any generation in our historical record," said the Rev. Gary Gunderson, the director of the interfaith health program at Emory University's public health school in Atlanta and an expert on AIDS.

Education can literally be a lifesaver for children who must grow up quickly—especially girls, whether by enabling them to learn how to make a living or to protect themselves against the virus, Mr. Gunderson added. "The crisis of HIV/AIDS," he contended, "has way more to do with people in schools than physicians in hospitals."

'Low-Grade Charity'

At the very least, some experts say, schools in sub-Saharan Africa could do a better job of helping children and teenagers avoid the virus. The infection rate among people younger than 15 is low, but then soars, especially for girls. So programs must start early—and go beyond simple awareness. In Cameroon, for example, more than nine in 10 teenagers were aware of AIDS, but fewer than 30 percent knew how to avoid contracting HIV.

Research strongly suggests that the lessons must impart both understanding of the disease and so-called "life skills" for risk management, such as negotiation with a sexual partner and identifying the ways in which a particular social environment poses dangers.

"So far, a lot of resources have been wasted in terms of money, material, and training that don't work," said Mr. Wright of UNICEF. "We're looking to revise [ineffective] programs."

The hurdles to pushing back the pandemic and minimizing its personal and national costs are high but not insurmountable, according to people such as Dr. Peter Piot, the UNAIDS executive director. Anti-AIDS efforts continue to be hampered by prejudice and discomfort because the disease is transmitted mostly by sex and results in death, according to Dr. Piot. He argues that

empowerment of people infected by HIV and those most vulnerable to it are among the best antidotes.

No one doubts that millions of Africans have mobilized to try to meet the needs, from compassionate groups operating on members' small contributions to education ministers convening policy and planning groups.

But inadequate funding must be reckoned with. Dr. Piot estimated that the $6.1 billion allocated for response to the pandemic in low- and middle-income countries in 2004 is half of what is needed for the current year. Moreover, the worst-affected African countries, he contended, would gain more from the cancellation of international debts, new trade practices, and better prices for pharmaceuticals than even from increases in rich nations' aid.

"The response already made to AIDS [by Africans] is quite unprecedented," said Mr. Gunderson of Emory. "What has not changed is our response [in the developed world]: We're acting as we've always acted, which is with relatively low-grade charity."

The public-health professor called on educators to add their voices to those of people living with AIDS and their advocates, who played an important role in putting drug treatment within the financial reach of more Africans. Educators in the United States and the rest of the developed world, he said, can grasp the challenges facing African children who have little and have lost much.

"This is an unprecedented crisis, in scale, and nature," Dr. Piot said in a speech in London last month, "and we have no choice but to act in exceptional ways."

UNIT 5

Population, Resources, and Socioeconomic Development

Unit Selections

Key Points to Consider

• How do you feel about the occurrence of starvation in developing world regions?

• What might it be like to migrate from your home to another country?

• In what forms is colonialism present today?

• For how long are world systems sustainable?

• Does it surprise you that Thomas Malthus is again in the news? Explain your position.

• Why is socioeconomic development in sub-Saharan Africa such a difficult task?

• What is your view on slavery in the 21st century?

• In your view, what are the five most serious world problems now?

Student Website

www.mhcls.com

Internet References

African Studies WWW (U.Penn)
http://www.sas.upenn.edu/African_Studies/AS.html

Geography and Socioeconomic Development
http://www.ksg.harvard.edu/cid/andes/Documents/ Background%20Papers/Geography&Socioeconomic%20 Development.pdf

Human Rights and Humanitarian Assistance
http://www.etown.edu/vl/humrts.html

Hypertext and Ethnography
http://www.umanitoba.ca/faculties/arts/anthropology/tutor/aaa_ presentation.new.html

Research and Reference (Library of Congress)
http://lcweb.loc.gov/rr/

Space Research Institute
http://arc.iki.rssi.ru/eng/

World Population and Demographic Data
http://geography.about.com/cs/worldpopulation/

© Goodshoot/Alamy

The final unit of this anthology includes discussions of several important problems facing humankind. Geographers are keenly aware of regional and global difficulties. It is hoped that their work with researchers from other academic disciplines and representatives of business and government will help bring about solutions to these serious problems.

Probably no single phenomenon has received as much attention in recent years as the so-called population explosion. World population continues to increase at unacceptably high rates. The problem is most severe in the less developed countries where, in some cases, populations are doubling in fewer than 20 years.

The human population of the world passed the 6 billion mark in 1999. It is anticipated that population increase will continue well into the twenty-first century, despite a slowing in the rate of population growth globally since the 1960s. The first article in this section suggests that, despite conflicts in the Middle East, the world is actually more peaceful than in past times. "Cloud, or Silver Lining" considers the socioeconomic consequences of an aging Japanese population. The next article discusses the continuing need for fresh water.

Thomas Malthus, whose historic theory is once more in vogue, is dealt with next. A review of programs to turn ocean water into tap water follows. The next article considers the prospects for containing the global spread of nuclear weapons. Finally, "A World Enslaved" shockingly states that 20 million people currently live in slavery in the world.

Wonderful World?

The Way We Live Now

Since the cold war, the earth has become more peaceful. Why doesn't it feel that way?

JAMES TRAUB

We live in a world that is objectively more dangerous than the one we knew at the outset of the millennium; whatever our disagreements, since 9/11 both our foreign policy and our domestic politics have pivoted around this ineluctable fact. And because the terrorists who struck us represent a global phenomenon, we assume that the world is objectively more dangerous for everyone. But is it?

The answer, it seems, is no. In recent years, scholars have been gathering data in an attempt to measure global trends in conflict and violence. The emerging view is that conflict worldwide is in fact diminishing, not growing. According to "Human Security Report 2005," a study that relies on this recent scholarship, the number of armed conflicts has been dropping steadily since the end of the cold war. Major civil wars, which exploded around the world between 1946 and 1991, declined sharply in the ensuing decade. Indeed, the trend holds true for virtually every category of conflict—coups, interstate wars and even genocides and so-called politicides, in which political belief rather than ethnicity is the criterion for killing. The only exception is international terrorism—as we know all too well.

The study, issued by the Human Security Center, a policy institute at the University of British Columbia, is not entirely convincing. Because, for example, no sound metric exists for tallying them, the "indirect deaths" caused by conflicts that uproot vast numbers of people, disrupt agriculture and shatter health-care systems have not been counted.

Nevertheless, once we get past our initial disbelief, the notion that we live in an era of comparative peace shouldn't be so surprising. Colonial struggles largely ended by the mid-70's. Cold-war proxy battles also ended. And over the last decade or so, the international community has developed a set of mechanisms to put the brakes on incipient warfare. In fact, the authors of the study argue that "the single most compelling explanation" for the decrease in conflict is that preventive diplomacy and peacemaking missions, as well as U.N. peacekeeping operations, which had been almost impossible to mount during the

cold war, now accompany almost every major crisis. As a result, outright warfare is now concentrated, albeit to an extraordinary degree, in one area—sub-Saharan Africa. Entire regions that were strife-torn 20 or 30 years ago, including East and Southeast Asia, are largely peaceful today. North Africa is quiet; Latin America suffers from political instability but not warfare.

This is certainly good news, especially for the beleaguered United Nations. But it does nothing to relieve our own sense of vulnerability, since terrorists are not amenable to negotiation or peacekeeping. You can, therefore, draw a troubling inference from the data, which is that for all the talk of "globalized" threats, the American experience of the world is becoming less, not more, similar to that of our allies and trading partners. Our world has become more dangerous; theirs, with some important exceptions, less so.

Why is it so hard for Americans to recognize how very different the world appears when seen from beyond our borders? The authors of "Human Security Report" hypothesize that myths about a rising tide of violence propagated by the media or international organizations "reinforce popular assumptions." O.K., but why the popular assumptions? Why are our intuitions about both the past and the present so far from the reality?

The answer has to do with fundamental differences between the cold war and the Age of Terror. In the cold war, the chief combatants never fought one another—thus the "coldness" of the war between them. The rest of the world was aflame with proxy combats, but the fear of provoking nuclear war served as a heat shield for the West. The authors of "Human Security Report" note that the scholarly description of the post-World War II era as the "long peace" is "deeply misleading." But psychologically it was accurate.

Now the relationship between threat and actual fatalities has been reversed. International terrorism has killed an average of 1,000 people a year over the last 30 years, according to "Human Security Report." That's not a lot. But the threat of terrorism is out of all proportion to its lethality; that is, in a way, the whole point of terrorism. It is true that since 9/11, nothing has happened in the United States, but anything can happen, anywhere, at any time. And of course something has happened. We all had nuclear nightmares in the depths of the cold war, but we recognized them as nightmares, or as scenes from apocalyptic movies. Yet in the case of terrorism, we had the apocalypse before we even had the nightmare. And so while we may accept the idea that we live in a less-war-torn age, the mind rebels at the thought that we live in a less dangerous age.

But that's us; it's not even "the West." In his famous essay "Of Paradise and Power," Robert Kagan argued that Europeans have stepped out of the Hobbesian world of anarchy—our world—into their own Kantian world of "perpetual peace." With Islamic terrorism a growing threat in Europe, Kagan's Mars/Venus distinction may not be quite as salient as it seemed.

But outside the West, where in general the world really does feel less lethal than it used to, President Bush has had a very hard time enlisting allies in the war on terror. Many developing nations would rather talk about development and trade than about sleeper cells. Yet we need their help in the fight against terrorism.

The Bush administration has discovered the hard way that diplomacy is more difficult today—and arguably more important—than it was during the cold war. You can only hope that this discovery points toward a wiser and more supple view of the world we have been forced to inhabit.

JAMES TRAUB, a contributing writer, is completing a book about Kofi Annan and the United Nations.

What Lies Beneath

As our nation seeks more ways to extract energy from sources outside the Middle East, the Bush administration is increasingly turning to oil, minerals, and natural gas that lie within and just beyond the borders of our national parks.

HEIDI RIDGLEY

As the afternoon sun casts a brilliant orange hue across the sandstone spires, a young girl and her father stop to rest, removing the burden of their backpacks, weighted down with water. The hike into the heart of Utah's red rock country wasn't particularly long, but it proved steep. Now nothing matters but the breathtaking view—until the child taps her father's arm and points at a strange figure in the distance, one that doesn't seem to belong.

On the horizon—shadowed, ominous, and unmistakable—stands an oil rig.

If the Bush administration continues its efforts to open federal land to energy extraction, such a scene could soon be commonplace in dozens of national parks. In Utah alone—a land of staggering beauty and sprawling, surreal landscapes—more than 172,000 acres faced possible leasing by the Bureau of Land Management (BLM) for energy production as of last August. Two of the parcels, roughly 3,200 acres, stood just four miles outside Canyonlands National Park—and within eyeshot of park visitors. Protests scuttled the plans. But representatives of the nonprofit Grand Canyon Trust are convinced that the BLM will propose the sales again and they resent the fact that existing leases on the land already border the park on two sides.

Parks that boast far greater visitation are also at risk. According to BLM's own statistics, air pollution from gas drilling in Wyoming's Powder River Basin—where rolling hills meet meandering streams—will hinder visibility at Yellowstone National Park to the west and Badlands National Park to the east. Haze from the drilling boom, expected to last two decades, is also expected to veil Mount Rushmore for more than 150 days a year.

But it's not simply an obscured view that concerns many conservationists. Studies show that besides denuding wilderness of vegetation to make way for roads and machinery, the industry's gargantuan footprint on the land can pollute groundwater and waterways and drastically disrupt wildlife migration for up to 80 miles. And the damage starts long before oil or gas is even found, as "thumper trucks" cruise the land, pulverizing the fragile landscape as they search for deposits with seismic waves that travel beneath the Earth's crust. Other times, exploration for oil and gas entails exploding dynamite in a hole drilled several hundred feet deep. If and when the industry strikes liquid gold, in come the heavy guns: 18-wheelers, diesel engines, and turbines—which often run 24 hours a day, seven days a week—and toxic, oil-based fluids used to keep the hole open and the drill bit cool.

Pinning its policy on the need to reduce the nation's dependence on foreign oil and enhance national security, the Bush administration has also worked steadily to whittle away at a long-standing principle to protect buffer zones just outside park boundaries from drilling and mining. According to a recent report by the Environmental Working Group (EWG), which analyzed well-by-well oil and gas production records obtained via the Freedom of Information Act, administration officials have removed barriers to drilling and lifted environmental protections on 45 million acres of public lands in 12 Western states.

According to EWG analyst Dusty Horwitt, 35 national parks in the region are at risk of environmental degradation due to mining, drilling, or both, and for no real reason. EWG's study found that from 1989 to 2003, the drilling on 229 million acres of federal land—the equivalent of more than Arizona, New Mexico, and Colorado combined—has provided the nation with only 53 days worth of oil and 221 days worth of natural gas. "Clearly, we're not going to drill our way to energy independence," says Horwitt. "Is it worth jeopardizing some of our most spectacular national treasures for a drop in the bucket?"

Farther east, another type of habitat destruction is endangering creatures protected under the Endangered Species Act while jeopardizing one of the best whitewater recreational rivers in the nation. At the headwaters of the Big South Fork of the Cumberland River in Tennessee, just outside the Big South Fork National River and Recreation Area, the coal industry is

Feeling the Impact

When a mining, oil, or gas corporation proposes to mine or drill on pristine public lands, the company often defends its development plans with statistics on the size of the operational footprint—the amount of land that will be denuded of vegetation for roads, buildings, concrete well pads, waste pits, processing facilities, and other infrastructure.

In reality, the footprint is rarely so remarkably contained as the industry would have officials believe: Government studies document that 40 percent of all Western headwaters are polluted with mine waste and that in some cases, plumes of smog rivaling those from big-city populations extend hundreds of miles from oil wells.

Crossing the Line: Drilling in Parks

Sadly, energy extraction doesn't always stop outside park gates. Because the process of creating a national park does not always include the transfer of mineral rights beneath the land, drilling platforms have cropped up inside a dozen of the nation's "crown jewels," and it's completely legal. Consider Padre Island National Seashore in Texas, the longest stretch of undeveloped barrier island in the world. Besides drawing almost a million visitors a year, a multitude of foraging shorebirds, 125 species of migrating neotropical birds, and the gulf's largest concentration of bottlenose dolphins, its dunes, mudflats, freshwater marshes, and beaches attract more than a dozen threatened and endangered species. One is the Kemp's ridley sea turtle, considered the most endangered sea turtle in the world.

To ensure that sea turtles don't get run over during drilling operations, the National Park Service mandates that the process be completed before nesting season begins in April. But with as many as 20 semi-trailers traversing the beach each day, the drilling operations not only disturb park visitors, they also compact the sand, says Carole Allen, Gulf Coast director for the Sea Turtle Restoration Project.

"Even if 18-wheelers don't run over adult turtles or hatchlings," says Allen, "they still pound the sand to a very hard surface, making digging a nest far more difficult." Drilling could also affect turtle mating or nesting. "No one really knows what effect these vibrations have on sea turtles," she says. "In 2004, 28 Kemp's ridleys nested in the area of Padre Island National Seashore. Would there have been more if there had been no drilling?"

gaining headway in mountaintop removal mining—a process that blasts away hilltops to get to the coal underneath. Traditionally, the mountaintops are left flat and the removed rubble gets dumped in nearby hollows or valleys. But these "fills" have caused so many environmental problems that the industry created a variation of the process—one that attempts to recreate the mountain. When mining operations cease in a particular area, the loose rock is piled back up—ostensibly in a way that ensures it won't slide down and pollute rivers and streams below. But that's proving to be nearly impossible in a steep-sloped area of the country that gets 50 inches of rain a year.

"It's like trying to pile oatmeal against a wall and expecting it to stay there," says Don Barger, senior director of NPCA's Southeast regional office. "It's a fantasy to think the rubble won't eventually slide down the mountainside and smother the aquatic life in the valley below."

While canoeing on the Big South Fork for several days last summer, Barger paddled by numerous sandbars that contained chunks of coal the size of his fist—the fallout from eroding mountain mining sites. Several endangered mussel species—filter feeders all—live in the Big South Fork, and they are simply unable to filter out all the silt caused by mountaintop mining.

Also affected by the removal of mountaintops is the cerulean warbler, a bird that has the unfortunate distinction of being the fastest declining warbler species in North America. Its primary habitat is mature forests on mountaintops above 2,000 feet, which makes coal mining its primary threat.

"The industry claims that the mountaintops can be replaced and revegetated with trees, and that this will solve the birds' problems," says Melinda Welton, a biologist studying the impact of mining on the birds in Tennessee. "The fact is that the geology of the mountaintops will forever be altered by mining. Ceruleans like the biggest, oldest trees, and we just don't know if it's possible to reestablish a forest that these birds will find acceptable—and we certainly won't know the results of this 'reforestation experiment' within our lifetime." While the warbler may be the poster child for the ill effects of mountaintop removal mining, a number of interior forest birds, such as the

Louisiana water thrush, the worm-eating warbler, and the wood thrush, are also seriously affected by coal mining.

Although the situation appears dire, it's not all doom and gloom. For the time being, at least, Glacier National Park in Montana stands as a prime example of what can happen if people band together and say no to irresponsible drilling. A paradise for more than 70 species of mammals and more than 260 bird species, Glacier comprises a million acres of forests, alpine meadows, and lakes. "This region is one of the most important intact natural areas in the world," says Steve Thompson, NPCA's Glacier program manager. "It's more intact than the Yellowstone ecosystem and Banff National Park in Canada. But the only reason that still rings true is because everyone got together to knock back the proposed oil and gas drilling on the fringes of the park: An Indian tribe, conservationists, hunters, ranchers, and small-town media stood up and said, 'Not here.'"

At issue are Glacier's sensitive species—the grizzlies, elk, and mountain goats that need large open spaces to thrive and survive. Putting in oil- and gas-drilling infrastructures would have meant roads that fragment habitat, which displaces wildlife and affects mating and foraging behavior, invites wildlife and vehicle collisions, and provides poachers greater and easier access to the wilderness. Unfortunately, the fight isn't over—Glacier borders British Columbia, Canada, where the government is more likely to "lease now and analyze the environmental effects later," says Thompson. Industry lobbyists also continue to push for new drilling on Montana's Rocky Mountain Front on Glacier's southern boundary. "Drilling is still a very serious lingering threat to Glacier, but for now, we've won."

But as Horwit notes, until our leaders realize that it's impossible to drill our way to energy independence, until they recognize that drilling in our natural treasures destroys the very things that make this country worth cherishing and defending, and until they see that the only way to reduce foreign oil dependence is to raise fuel economy in cars and embrace alternative energy, the fight will never really be over.

HEIDI RIDGLEY is a freelance writer living in Washington, D.C.

Cloud, or Silver Linings?

Japan's population is ageing fast and shrinking. That has implications for every institution, and may even decide the fate of governments.

For intriguing evidence of the way Japan's 127m people are greying faster than any others on earth, look at the boom in *pokkuri dera. Pokkuri* is an onomatopoeic word for a sudden bursting, while a *tera* or *dera* is a Buddhist temple. *Pokkuri dera,* then, are shrines where many of Japan's older people go to pray not only for the long life that they are increasingly coming to expect, but also for a quick and painless death at the end of it. Their visits have revived the fortunes of old-established temples, notably in the ancient capitals of Kyoto and Nara, while temples elsewhere have reinvented themselves as *pokkuri dera* with the financial blessings in mind.

More dramatic evidence of the ageing effect may come with nationwide elections on July 29th. Japan's older voters have the ability, for probably the first time in democratic history, to humiliate and even bring down a government, that of Shinzo Abe, prime minister since September 2006. The elections are for half the seats in the upper house of the Diet (parliament), and are ordinarily something of a political sideshow: after all, it is the lower house that chooses the prime minister. A general election in 2005 gave the ruling coalition led by the Liberal Democratic Party (LDP) an easy majority. Yet these elections, in which the coalition may lose its upper-house majority in the lower house, have become a vote of confidence in Mr Abe, whose poll ratings have slithered since almost the moment he came to office.

While the prime minister's priorities are patriotic ones—instilling a sense of national pride in schoolchildren and pushing for a revision of Japan's pacifist constitution—those of ordinary Japanese lie with bread-and-butter issues. The economy is now into its fifth year of recovery after a decade-long slump, but decent jobs are still short. As for pensions, everyone knows that a shrinking workforce supporting an ever higher number of retired people adds to an already strained budget.

In this context, a fiasco that was uncovered in May at the government agency that handles pensions could not have come at a worse time for Mr Abe. The agency, which appears never to have come to terms with the digital age, is unable to match 50m computerised pension records to people who have paid into public schemes. A further 14m records, it seems, never made it into the computer system at all.

If disgruntled voters punish the ruling coalition on July 29th with a heavy loss of seats, then the LDP may seek a new leader.

If Mr Abe survives as prime minister, he will be under pressure to form a government of a different hue, one that brings livelihood issues to the fore. Either way, grey power will have established itself as a force to be reckoned with.

Certainly Japan is greying at an astonishing rate. Shortly after the second world war the proportion of Japanese over 65 was around 5% of the population, easily below that in Britain, France or America. Today the elderly account for one-fifth of the population, and average lifespans have grown remarkably. Life expectancy today is 82, up from a little over 50 in 1947.

By 2015 the proportion of elderly will have risen to one in four of the population, or more than 30m. This is thanks mainly to an unusually large baby-boom generation passing into the ranks of the old. Between 1947 and 1949, 2.7m children a year on average were born to surviving Japanese soldiers who returned from war, married and settled down—about a third more than in previous years. This year, the baby-boom generation began to retire (at present, 60 is the mandatory retirement age at most companies). The size of their pensions obligations has funding implications both for companies and for government. But there is another dimension to the baby-boomers' retirement: these workers drove Japan's economic transformation of the 1970s and 1980s. They are a reservoir of technical and managerial skills.

Who to pass these on to? Japan's birth rate fell below the replacement rate of 2.1 in the early 1970s. It slid to a low of 1.26 in 2005, before inching up last year to 1.32—nobody calls it a recovery. In 2005 Japan's population began to fall in absolute terms, despite increasing life expectancy. It is about to shrink at a pace unprecedented for any nation in peacetime. The National Institute of Population and Social Security Research estimates a total population of 95m by 2050, with the elderly accounting by then for two-fifths of the total.

The Disappearing Young

A shrinking population already has implications for the workforce. Currently, some 16m Japanese are in their 20s. This number will shrink by 3m over just the next decade. This spring, during the annual job-recruitment round, new university graduates found themselves in record demand, and not just because of the recovering economy: over the coming years, companies will have fewer young graduates to choose from. That is nice for young job-seekers, except for one thing: as Japan ages and shrinks, workers must support an ever larger proportion of retirees. By 2030, demographers say, Japan will have just two working-age people for each retired one; by mid-century, short of a rapid and unlikely return to fecundity, the ratio will rise to three for every two retirees.

Can a working population support such a number of future retirees? Today's younger workers appear not to think so. Two-fifths of them are not paying contributions towards the fixed portion of their state pension scheme (current contributions fund present, not future retirees), suggesting they don't believe that the scheme will be viable when they retire. And they may be right.

It is in the countryside that demographic changes hit particularly hard. There the population has been falling for years, as younger villagers head for the city in search of work and play. Today, those over 65 account for two out of five people in rural communities, and three-fifths of all farmers. The future of farming in such places is in doubt. Growing rice, the staple crop, requires communal efforts in irrigation, flood control and the like. Mutual obligations in communities run even to organising funerals. So when young villagers leave for the city, everyone feels the loss. An earthquake on July 16th in Niigata prefecture brought the problem home; the 3,000 evacuees still living in shelters are predominantly elderly, unable to fend for themselves in their damaged houses.

The tiny hamlet of Ogama, in Ishikawa prefecture near the Sea of Japan, is responding most radically to population decline. (The community has three men and six women between the ages of 62 and over 90, down from a population of 50 a generation ago.) The survivors of this remote and stunning valley have canvassed an industrial-waste company from Tokyo and, if the prefecture approves, the valley—paddy fields, vegetable plots and cedarwood plantations—will disappear under 150 metres (500 feet) of industrial ash. The villagers plan to use the money from the sale to build new houses in the nearby township, to where the ancestral shrine has already been moved.

For years, the regions have brought their problems to the capital. On any working day in Tokyo, the corridors of the transport and infrastructure ministry are thronged with supplicants from the provinces clutching maps of the latest scheme for a road into the forest or an unnecessary dam. Yet the days of lavish spending on public works are nearly over, while the central government has slashed tax remittances to localities. With pinched resources and the prospect of steep falls in the population, local governments are being forced into the most radical reorganisation in half a century.

A couple of much-publicised municipal bankruptcies have helped sharpen minds. Yubari, a former mining town on the northern island of Hokkaido, has seen its population fall from 100,000 in the 1950s to 13,000 today. Costly promotions to raise the town's profile—including a film festival and the marketing of Japan's priciest melons—have saddled the town with a crippling ¥63 billion ($519m) in debts. Last year Yubari was declared insolvent.

No nearby municipalities particularly want to be Yubari's friend, but elsewhere the central government is urging villages and towns to merge in order to pool resources and gain a more secure tax base. Yamanashi prefecture south-west of Tokyo, a place of peach orchards and factories making industrial robots, exemplifies the trend. In 1888 Yamanashi had 342 administrative units; today, it has shrunk to 28 municipalities and is still declining. The pace has quickened greatly since 2003.

But municipal mergers are unlikely to be the end of the matter. Prefectural leaders and central government are talking about a radical rehaul of local government in which prefectures merge to form larger blocks—states, in essence. Before this dance has begun, prefectures are already eyeing up the most attractive partners.

To Shrink a City

Elsewhere, administrators are starting to think about the implications of population decline, among other things, on running bigger cities. Aomori, a city of 300,000 at the very top of Honshu, Japan's main island, has a policy of actively stemming the urban sprawl that blights so much of Japan. Aomori has a proportion of elderly and single households somewhat above the national average. It also has huge quantities of winter snow, thanks to the moisture that Siberian winds pick up across the Sea of Japan: ten metres can fall in a season. In a bad year snow-clearing can cost ¥3 billion: a sum which Takeshi Nakamura of the city government says could build two new schools.

In response, the municipal government set about trying to shrink the city. A limiting arc was drawn around its south side (the north is bounded by a wide bay), and some of the city's main institutions—the library, city market, hospitals and museums—were moved back to the middle of town. Public transport was improved, and snow was cleared from main arteries as well as pedestrian streets to allow people to move easily about the centre. The improvements, in turn, have encouraged new apartment blocks to be built near the centre, says Mr Nakamura, and plenty of older people tired of shovelling snow are moving into them.

Aomori's ideas about a "compact city" have been driven by the problems of snow. All the same, says Takatoshi Ito of Tokyo University, who sits on Mr Abe's Council for Economic and Fiscal Policy, the central government should be urging other cities to think along similar lines. Population decline does not mean there is no urban sprawl. Mariko Fujiwara of the Hakuhodo Institute of Life and Living points out that the number of one-person households will overtake all other types this year, while the total number of households is still rising in Japan, to almost 50m.

Kaisha Care

Still, the greatest response to demographic change in Japan needs to come from companies. Despite wrenching change over the past 15 years or so, the Japanese company, or *kaisha*, still plays a more paternal part in employees' lives than in any other well-off society, shaping not just their work but also their social life. Indeed, with long hours in the office as well as punishing sessions in bars with colleagues afterwards, the two are often indistinguishable. Atsushi Seike, a labour economist at Keio University, argues that Japan's problem is less that demography is changing too fast, than that employment and retirement systems designed for an earlier age are not changing fast enough.

In particular, these systems have not kept pace with greatly longer lives. True, the government has begun to raise the age at which people are eligible for employee pensions, which are made up of fixed and earnings-related parts. Eligibility for the fixed part has been raised to 62, and will climb to 65 by 2014; eligibility for the bigger, earnings-related part rises to 65 by 2026. This is too little, too slow. Mr Seike argues that the state minimum pensionable age should be raised swiftly to 70.

Meanwhile, companies are also adjusting too slowly. Most firms have a mandatory retirement age of just 60. A recent law requires them either to raise their mandatory retirement age over time, or to provide retraining and re-employment programmes to keep on employees. Most have opted for the latter; since most companies have formal pay scales that reward seniority over merit, raising the mandatory retirement age would be expensive. However, one big company, Kawasaki Heavy Industries, has broken new ground: in 2009, it will raise mandatory retirement to 63 while slashing pay.

Getting rid of mandatory retirement altogether would hasten the end of seniority-based pay, allowing older workers (who in Japan are eager to work for longer) to fill jobs for which they are best suited. A system based more on merit would give able younger workers a leg-up too.

Raising the retirement age to 70 would roughly halve the rate of decline in the workforce. Raising the participation rate of women—at 63% of working-age women, below Britain or America (around 68%)—would do much to slow it further. A number of factors militate against working women. A higher proportion of women than men find jobs only on temporary contracts, which pay on average 60% less than regular work. Male chauvinism still dominates in the office: many jobs are advertised as available only to younger women, while fewer than 10% of professional managers are women, against 46% in America. Meanwhile, companies' long hours (often a substitute for productivity) make things hard for working mothers. So too does a shortage of child care: just a third of children over three and under school age go to kindergarten, compared with an OECD average of three-quarters. Huge numbers of women drop out of the workforce entirely once they have children. In Japan, says Jeff Kingston of Temple University in Tokyo, women have to choose between work and family.

Meanwhile, the OECD notes a positive correlation between fertility and female employment: the easier it is made for women to do rewarding work, the more likely they are to consider having children. So policymakers in Japan are now starting to grapple with the effect of Japanese work habits on the low birth rate. Hideki Yamada, director for policy on ageing and fertility in the Cabinet Office, says surveys suggest that nine-tenths of Japanese aged 18-34 not only want to get married, but often want to have two children. With Japanese precision, policymakers have calculated that without impediments to marriage and child-raising, Japan's birth rate would jump to 1.75.

Policy, says Mr Yamada, should be directed towards making that leap. Attempts began under Mr Abe's predecessor, Junichiro Koizumi, with the introduction of financial support for families with young children and expansion of child-care facilities. Now a novel concept is creeping into government documents, "the work-life balance," for which, tellingly, there is no common Japanese expression. In late July, business and union leaders met Mr Abe and other ministers to discuss how to reach such a balance.

"It's embarrassing to say this," admits Mr Yamada, "but after a first child is born, the husband often doesn't do his bit helping out at home, and that engenders anxiety in the wife about having a second child." That is partly cultural habit. Boys are pampered at home by their mothers and expect the same treatment—no nappy-changing, no washing up—later from their wives. But it is also because of the long working hours companies expect. So, says Kuniko Inoguchi, minister for gender issues and social affairs under Mr Koizumi, policy needs not only to be directed towards encouraging more women to work, with more nursing care for elderly relatives, better child care, more flexible working arrangements and so on. It also needs to make life better for working men.

A better work-life balance is good for companies, which can thereby attract better talent. It is also good for working men, says Mrs Inoguchi. They can enjoy a proper private life, spending more time at home—always assuming, and it is no foregone conclusion, that Japanese wives are prepared to tolerate them there.

Troubled Waters

Drought, pollution, mismanagement, and politics have made water a precious commodity in much of the world.

MARY CARMICHAEL

Daily life in the developed world has depended so much, for so long, on clean water that it is sometimes easy to forget how precious a commodity water is. The average American citizen doesn't have to work for his water; he has only to turn on the tap. But in much of the rest of the world, it isn't that simple. More than a billion people worldwide lack clean water, most of them in developing countries. The least fortunate may devote whole days to finding some.

When they fail—and they fail more and more often now that rivers in Africa and Asia are slowly drying up after decades of mismanagement and climate change—they may turn to violence, fighting over the small amount that is left. Water has long been called the ultimate renewable resource. But as Fred Pearce writes in his book *When the Rivers Run Dry*, if the world doesn't change, that saying may no longer apply.

Like the famines of the '80s, the global water crisis is far more than a straightforward issue of scarcity. Accidents of geography, forces of industry and the machinations of politics may all play a role in who gets water—just as warlords, as well as droughts, were responsible for starvation in Ethiopia. In many ways, the famines contributed to today's man-made droughts: the crops grown in the worldwide "green revolution" of the past three decades sated hunger but sapped water in the process. "As the globe gets more crowded," says Susan Cozzens, a policy professor at Georgia Tech who is working on water problems, "the old arrangements just don't work anymore."

There is still time for nonprofits and governments to fix things. "Chlorination, gravity-fed distribution systems, taps at every household, all these could make a difference," says John Kayser of Water for People, a nonprofit working in the developing world. Ecoconscious start-ups in the United States and Europe are increasingly offering new ways of purifying water, from high-tech (but inexpensive) ultraviolet filters to simple tactics such as filling clear bottles and letting the hot sun kill the bacteria inside.

But thus far, there has been no worldwide "blue revolution." More likely, says Pearce, we'll "only really start to worry about the water when it isn't there." Here are some flashpoints, regions where the future of water is most worrisome.

China's Poisoned Water

To look at the mighty Yangtze River, you might think China could not have a water crisis. The third longest river in the world, it funnels 8 million gallons into the East China Sea every second. The river drives the world's largest hydroelectric dam, the Three Gorges, and it is one of the backbones of the country's economy.

When you look more deeply into China's water supply, however, you'll see plenty to worry about. The government has long known that the Yangtze is polluted. In 2002, Beijing announced a $5 billion cleanup effort, but last year admitted that the river was still so burdened with agricultural and industrial waste that by 2011 it may be unable to sustain marine life, much less human life. An April report by the World Wildlife Fund and two Chinese agencies found that damage to the river's ecosystem is largely irreversible.

Travel farther north, especially near the country's other major water system, the Yellow River, and the picture is even bleaker. Since the 1980s, drought and overuse have diminished the river to a relative trickle. Most of the year, little to none of its water reaches the sea, says Pearce. What does still flow in the Yellow is often unsuitable for drinking, fishing, swimming or any other form of human use. Every day, the river absorbs 1 million tons of untreated sewage from the city of Xian alone.

Nowhere is China's pollution problem more visible than in the tiny "cancer villages" that dot the country's interior. Shangba, a town of 3,000, captured national attention a few years ago after tests found that heavy metals in its local river far exceeded government levels. Officials from a nearby state-owned mine—suspected of dumping those chemicals into the water—persuaded the government to pay for a new reservoir and water system built by locals. But other, smaller cancer villages are still struggling. In the southern hamlet of Liangqiao, rice grown by villagers with water from a local river has taken on the reddish hue of contaminants from the same iron mine that blighted Shangba. Since the late 1990s, cancer has caused about two thirds of the 26 deaths in the village. "We have to use the polluted water to irrigate the fields, since we have no other choices. We don't have any money to start a water project," says Liangqiao resident He Chunxiang. "We know very well that we are being poisoned by eating the grain. What more can we do? We can't just wait to starve to death."

There is hope yet for Liangqiao. Environmental lawyer Zhang Jingjing is filing a lawsuit against the mine on behalf of the villagers, and she has a strategy that focuses on loss of crops instead of loss of life. (Chinese courts are often reluctant to link cancer to pollution.) But win or lose, Liangqiao is a tiny part of the problem. It has just 320 people. Meanwhile, almost 400 million Chinese, fully a third of the country's population, still have no access to water that is clean enough for regular use.

India's "Hydrological Suicide"

In this country of 1.1 billion, two thirds lack clean water. "Sanitation for drinking water is a low priority there, politically," says Susan Egan Keane of the Natural Resources Defense Council. The priority is agriculture. In

Not a Drop to Drink

Industrialized nations not only have greater access to clean water than the developing world, they also tend to use more and pay less for it. An overview:

–MARC BAIN

573 LITERS: Average water use per person per day in the United States. In Italy, it is 385; in China, it is 87; and in Mozambique, it is 10 liters.

% OF POPULATION WITH SAFE DRINKING WATER

- ■ 100%
- ■ 90–99
- ■ 80–89
- ■ 70–79
- ■ 60–69
- 50–59
- 40–49
- ■ 39 or less
- ■ No data

1.8 MILLION: Number of children who die every year from diarrheal diseases (including cholera); 41% of those deaths occur in sub-Saharan Africa.

70 CENTS: Cost of a cubic meter of water in New York City in 2003. In London, it was $1.80; in Manila, $2.92; in Barranquilla, Colombia, it was $5.50.

the '70s and early '80s, the Indian government made this clear by pouring money into massive dams meant to pool water reserved for farms. "In many of these developing countries, the vast majority of their fresh water goes to irrigation for crops," says Egan Keane. "Agriculture may make up only 25 percent of the GDP, but it can get up to 90 percent of the water."

That's not to say, however, that India's farmers have enough. They are actually running low. The government built dams, but it failed to create the additional infrastructure for carrying water throughout the countryside. At the same time, factories have drawn too heavily on both the rivers and the groundwater. In Kerala, a Coca-Cola plant had to be shut down in 2004 because it had taken so much groundwater that villagers nearby were left with almost none.

Some farmers have reacted wisely to the dropping water levels by switching to hardier crops. Kantibhai Patel says he stopped growing wheat on his farm in Gujarat, the epicenter of India's water shortage, after eight years of watching his bounty and income shrivel in the sun. He farms pomegranates now, which require far less water than wheat. Experts hope more farmers will follow Patel's lead. So far, most farms still focus on water-guzzling crops like wheat, cotton and sugar cane. Indian dairy farmers also cultivate alfalfa, a particularly thirsty plant, to feed their cows, a practice Pearce calls "hydrological suicide." For every liter of milk the farmers produce in the desert, they consume 300 liters of water, says Saniv Phansalkar, a scholar at the International Water Management Institute. "But who is going to ask them not to earn their livelihood," he asks, if the dairy farms are keeping them afloat for now?

To nourish their plants and cows, most Indian farmers have resorted to drawing up groundwater from their backyards with inexpensive pumps. When the pumps don't bring up enough water, the farmers bring in professionals who bore deeper into the ground. There is constant pressure to compete. "If one [farmer] is digging 400 feet into the ground, his fellow farmer is digging at least 600 feet," says Kuppannan Palanisami, who studies the problem at Tamil Nadu Agricultural University. The water table, he says, drops six to 10 feet each year.

The West Bank's Water Wars

Like the Chinese, the people of the West Bank wouldn't have a huge water problem if nature were the only force involved. Rain falls regularly on their hills and trickles down into the rocks, creating underground reservoirs. Unfortunately for denizens of the West Bank, that water then flows west toward Israel. Palestinians are largely banned from sinking new wells and boreholes to collect water, and they pay what they consider inflated prices to buy it. Meanwhile, the groundwater level is dropping, and Palestinians accuse Israelis of overusing. "[The Palestinians] sit in their villages, very short on water," says Pearce, "and they look up at their neighbors and see them sprinkling it on their lawns."

Battling over water in this region is nothing new—the Six Day War started with a dispute over water in the Jordan. Lately, on the West Bank, the water table is dropping and tensions are rising. Israeli soldiers have been accused of shooting up water tanks on the West Bank in retribution for terrorist acts, and Palestinians have been caught stealing from Israeli wells.

It is impossible to untangle the water problem in Israel and the Palestinian Authority from the overall animosity between the two groups. Conversely, says Pearce, "the wider problem between the Palestinians and the Israelis won't be solved until the water problem is solved." On the West Bank, it's a Catch-22, with water—and life—on the line.

With Sarah Schafer, in Beijing and Sudip Mazumdar, in New Delhi

Turning Oceans into Tap Water

Desalination promises to rescue sprawling communities in dire need of freshwater. Is that a good idea?

TED LEVIN

America is running out of drinking water. In parts of the arid West, this is literally true. In coastal areas, such as Pinellas County, Florida, the problem more closely resembles Coleridge's famous verse, "Water, water, every where/ Nor any drop to drink." To slake its thirst, the local water authority, Tampa Bay Water, has built the largest desalination facility this side of Saudi Arabia. Situated on Apollo Beach, just across Tampa Bay from the Pinellas Peninsula, the plant is the only operational commercial desal facility in the United States. Eventually it will supply the region—a three-county area with more than two million people and growing—with 10 percent of its drinking water. (The rest will come from a now depleted aquifer, a new groundwater supply, and several aboveground rivers.)

The Apollo Beach plant may be a very good idea or a very bad one. It all comes down to this: Is desalination a legitimate response to a bona fide emergency, or is it simply an enabler for unchecked sprawl in fragile coastal areas that do not have the natural means to support their exploding populations?

Pinellas County, home of lovely St. Petersburg, is bounded on the west by the Gulf of Mexico and on the south and east by Tampa Bay. The soil is sandy and porous, perfectly suited for the engineering works of gopher tortoises. The beaches are classic Florida, bone-white sand lapped by blue water, beneath a wide arc of subtropical sky. In 1539, when Hernando de Soto marched up the Gulf coast, the Pinellas Peninsula was an open woodland of pines and palms and oaks. A dense coif of mangroves punctuated by salt marshes rimmed Tampa Bay, while the bay itself, covering nearly 400 square miles, was a mosaic of sea grass beds and oyster bars, mudflats and open water. In season, birds from across the continent convened in and around Tampa Bay to gorge themselves on the flats and beaches and in the woodlands and shallows, where shoals of fish moved from the Gulf to spawn or feed in the fecund estuarial waters. Sea turtles nested on the beaches. Manatees grazed the sea grass beds. Back then, before the dredging of shipping lanes, a man could have threaded his way across the shallow bay without wetting his hair.

Tampa Bay remained a symphonic wilderness well into the nineteenth century, but its despoliation was swift. In the late 1880s, the hub of Pinellas County was an unnamed community, population 30. In 1892, the community incorporated into St. Petersburg, population 400. Early in the last century, to meet future water needs, Pinellas County and the city of St. Petersburg bought land in the hinterlands of Pasco and Hillsborough counties, north of Tampa Bay. Eleven well fields set in remote wetlands supplied the city with the potable ground-water that the peninsula itself could not provide.

By 1920, the population of Pinellas County had reached 28,000. Five years later, after a six-mile bridge was built to connect Pinellas County and Tampa, the population had grown to 50,000. By 1950, it was 159,000. By 1970, it had soared to 522,000. Today, as Pinellas County's population reaches nearly a million, Pasco and Hillsborough counties have undergone population explosions of their own, further stressing the well fields. Surrounding wetlands have become fire hazards and nearby lakes have receded from their shores. The faucets of some Pasco County residents literally have run dry.

A century of dredging, filling, building, and digging has destroyed 80 percent of the sea grass beds and more than 40 percent of the mangroves and salt marshes. Storm water run-off from cities and farms and the dumping of untreated sewage continue to strangle Tampa Bay. Nitrogenous compounds from coal-fired power plants and automobile exhaust fall out of the air, lacing the rain with toxins and turning the bay's gin-clear water into an opaque algal soup that has smothered the sea grass beds.

Only 3 percent of the earth's water is fresh, and more than two-thirds of that is bound up in glaciers and ice caps, rock-hard and beyond reach. This leaves less than 1 percent of the planet's water available for drinking and washing and mixing with bourbon, and that meager amount is not evenly distributed.

On the face of it, the Tampa Bay region would seem to have an abundance of aqueous resources. Buried among the layers of sedimentary rock beneath Florida and its continental shelf lies an ancient bubble of freshwater, the Floridan Aquifer, one of the largest in the world. Like the state, the aquifer is bounded on three sides by salt water. The layered rocks hold roughly two quadrillion (that's 2,000,000,000,000,000) gallons of water. To this hefty amount add 50,000 miles of rivers and streams, nearly 8,000 lakes and ponds, and 600 springs, some so large they become navigable rivers when they reach the surface. All this water sits on, or under, or slices through, more than three million acres of wetlands. When compared to other Sun Belt states, Florida appears submerged in good fortune. The question arises, then: Why are the 11 well fields that serve the greater Tampa Bay area running out of water?

One reason is that groundwater does not behave like surface water. Wells take longer than lakes to recharge, and the lower pressure created by depleted wells pulls surface water downward. The more water drawn out of a well field, the deeper and wider the zone of lower pressure, and the more surface water fills the void. As surface water drains away, wetlands dry out, and even though particular localities sit atop a subterranean sea of freshwater, they may suffer a dramatic loss.

Prior to the passage of the state's 1972 Water Resources Act, which established five regional water management districts within the Florida Department of Environmental Protection, anyone could drill anywhere. After 1972, the water management districts began to issue consumptive use permits. Twenty years later, when Pinellas County's groundwater permits expired and Pasco County balked at having them renewed, the crisis moved from the faucets to the courts, eating up millions of dollars in legal fees.

In 1997, after a lengthy and contentious review process, the Southwest Florida Water Management District agreed to cofund a search for new supplies of freshwater for the Tampa Bay area. In an effort to alleviate Pasco County's water shortage, the water management district agreed to scale back pumping of the well fields. The goal was to reduce the level of pumping by more than half—from 192 million gallons a day (mgd) in 1996 to an eventual low of 90 mgd by 2008. This reduction, hydrologists hoped, would be enough to restore the health of the aquifer. By 1998, continued water shortages forced the governments of Hillsborough County, Pasco County, Pinellas County, St. Petersburg, New Port Richey, and Tampa to try something new. They decided to commission the construction of what would be the largest desalination plant in the country.

Until very recently, the notion of drinking seawater was lunatic fringe, involving a technology suitable for nuclear submarines and the Middle East, where an oil-rich, water-poor landscape makes financial and practical obstacles irrelevant. In 1960, there were just five desalination plants worldwide. Until the late 1990s, only two American cities had invested in full-fledged desal plants—Key West, Florida, in the 1980s, and Santa Barbara, California, a decade later. Both cities shelved their plans soon after the facilities were built, having found less expensive sources of water elsewhere. It is still cheaper for Key West to pump freshwater 130 miles from beneath the apron of the Everglades than to desalinate seawater.

However, as desalination technology improves, lowering the cost of producing freshwater, more planners are looking to the ocean as the droughtproof guarantor of continued growth. Throughout the Sun Belt, metastasizing communities have outstripped existing water supplies and begun to look seaward. Last year, municipal water agencies from California, Arizona, New Mexico, Texas, and Florida pooled resources and formed the U.S. Desalination Coalition, a Washington, D.C.-based advocacy group that lobbies the federal government to invest in new desalination projects.

Today there are more than 12,500 desal plants in 120 countries, mostly in the Middle East and Caribbean. Saudi Arabia meets 70 percent of its water needs by distilling salt water; the British Virgin Islands Tortola and Virgin Gorda rely on desalination for 100 percent and 90 percent of their respective water needs. The American Water Works Association, the largest organization of water professionals in the world—its 4,500 utility members serve 80 percent of America's population—forecasts that the world market for desalinated water will grow by more than $70 billion in the next 20 years.

California will soon be in the vanguard in the United States. It has already planned or proposed about a dozen desal plants along its coast, including a $270 million plant in northern San Diego County slated for completion in 2007. Early last year, the federal government reduced the amount of Colorado River water allocated to Southern California, forcing the state to accelerate its search for alternative sources after years of helping itself to the dun-colored Colorado at the expense of other western states (and Mexico).

To learn about the potential impact of desalination, I visit Mark Luther at the University of South Florida's Marine Science Center, in St. Petersburg. After a slow drive across the Pinellas Peninsula, traffic congealing at every intersection, I pull into the science center parking lot. It's an early December afternoon, hot and dry, the sky blue from seam to seam. High above the lot, an osprey throws a tantrum, lobbying for issues beyond my comprehension. From the second floor of the building I can see the desal plant across the bay on Apollo Beach, white like the salt it removes. Luther is the oceanographer who studied the bay's circulation patterns as part of the environmental assessment team that helped Tampa Bay Water determine where to site the facility. We settle at a black laboratory table in his bright, cluttered office. Luther, 50, wears a powder-blue yacht club T-shirt and sockless moccasins. His eyes match his shirt. His sand-colored, shoulder-length hair hangs in a ponytail. Luther tells me that, on average, 60 cubic meters of freshwater a second flow into the head of Tampa Bay, courtesy of four main rivers—the Hillsborough, the Alafia, the Manatee, the Little Manatee—and a number of smaller tributaries. The freshwater, lighter than salt water, is stirred by the tides before draining into the Gulf of Mexico.

"No matter where you take freshwater, it's going to have some impact on the environment," Luther says. "The goal is to distribute the sources to reduce that impact." Besides operating the desal plant, Tampa Bay Water pumps two new groundwater sites and diverts water from three of the rivers that feed Tampa Bay. "Taking river water has a much larger impact on the bay than the desalination facility," he says. "Of all the ways to get potable freshwater, building a desal plant is no worse and probably better than overpumping well fields or diverting too much river water." It's hardly a ringing endorsement, but it also suggests that an intelligently planned desal plant is not something a sensible environmentalist should lose too much sleep over.

You can't locate a desalination plant just anywhere, however. You need an energy source to operate the plant and a circulation pattern that removes the discharged brine. Brackish water, being less salty than seawater, costs less to desalinate. Hence, the plant was built inside the bay, on Apollo Beach, where salinity, though varying seasonally, averages 20 parts per thousand (ppt), 15 ppt lower than in the Gulf of Mexico. The Big Bend coal-fired power plant sits next door, providing a ready source of water and energy. Of the 1.4 billion gallons the power plant uses each day to cool its condensers, Tampa Bay Water recycles 44 million gallons for desalination. Because the plant already passes intake water through a pair of screens to filter out fish and other sea organisms, from fish eggs to plankton, the desal facility does not cause any additional loss of aquatic life. From the 44 million gallons of salt water it receives daily, the plant produces 25 million gallons of freshwater. The highly concentrated salt water that remains is mixed with the power plant's effluent before being returned to Tampa Bay.

This discharge water adds only marginally to the salinity of the bay, says Luther. A little more than a quarter of a mile from the discharge site he could not detect any increase in salinity. "We're at least an order of magnitude less than natural variability," he reports. The circulating currents and tides, aided by a 43-foot-deep shipping lane dredged decades ago, wash the brine away from Apollo Beach.

Not everyone believes the desal plant is benign. According to an advocacy group called Save Our Bays, Air and Canals (SOBAC), which has its headquarters in Apollo Beach, Tampa Bay takes two years to flush. The briny discharge, SOBAC claims, is equivalent to dumping a truckload of salt in the bay every 16 minutes. The group says that part of the littoral zone off Apollo Beach is already hypersaline. Luther does not believe the desal plant will add to the problem. This part of Tampa Bay flushes about every two weeks during the summer, he tells me, less frequently during the winter. "The waters off Apollo Beach are constantly refreshed. That's why the site was chosen.

"It's ironic that SOBAC brings up hypersalinity," Luther adds. "Probably the biggest environmental disaster to hit Tampa Bay in the last 50 years was the construction of the Apollo Beach community. They dredged pristine mangroves and sea grass beds to build stagnant finger canals and spits of land that

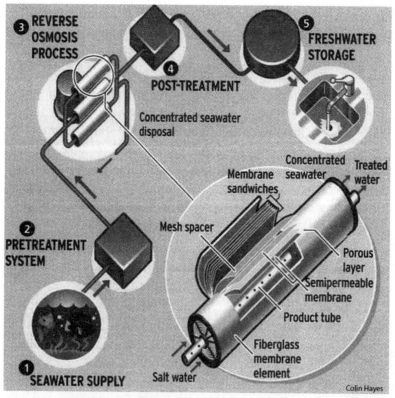

Put simply, desalination purifies water by removing dissolved mineral salts and other solids. In the Middle East, most desalted water is produced by means of distillation, which imitates the natural water cycle: Saltwater is heated to produce water vapor, which then condenses to form freshwater. American desal plants favor a different technology—reverse osmosis—which forces the water through a series of membranes, leaving the salts behind. Tampa Bay Water engineer Ken Herd, top, shows a cutaway model of one section of wound membrane. Hollow at its core, each section conveys the desalination water to the final "post-treatment" phase.

are now heavily developed. All those waterfront homes have nice green sodded lawns that require fertilizers and pesticides, which drain right into Tampa Bay."

As a naturalist, I know that filtering salt from seawater is not a novel idea. For hundreds of millions of years marine plants and animals have evolved unique methods of desalination. Salt glands discharge excess salt through the nostrils of marine iguanas, the eyes of sea turtles, and the tongues of crocodiles. The underside of the leaves of black mangrove trees exude pure salt crystals that glisten in the tropical sun; the spidery roots of red mangroves block salt from entering the tree. The gills of saltwater bony fish such as tuna or striped bass, the rectal glands of sharks and rays, and the super-kidneys of whales and seals perform a similar function.

I want to understand how desal works for humans, so I drop in on Ken Herd, 43, engineering and projects manager at Tampa Bay Water's Clearwater office complex. Tampa Bay uses a reverse osmosis (R.O.) membrane system, explains Herd, in which salt water is pushed at extreme pressure, up to a thousand pounds per square inch, through tiny pores, each 0.0001 micron in diameter—approximately 1/1,000,000 the width of a human hair.

Osmosis, as you may recall from 10th-grade biology, is the tendency of a fluid to pass through a semipermeable membrane, such as the wall of a living cell, into a solution of higher concentration, to equalize concentrations on both sides of the membrane. Reverse osmosis is precisely . . . the reverse. The

pores of the roughly 10,000 tightly rolled membranes are so small that ultratiny molecules of water pass through, but larger molecules of dissolved minerals like salt do not. Pressure forces out the salt, and the constant flow of water helps wash the outer membranes clean of concentrations of brine. R.O. membranes still clog, however, and have to be cleaned, every three weeks to six months or longer. The membranes last five to seven years, sometimes ten, and they are expensive to replace.

Herd shows me a model of a three-foot section of wound membrane. It looks like an oversize roll of paper towels, with the top cut away so that I can see inside. Salt water forced against the outside of the roll filters through the spiral until pure freshwater flows into the center port—the equivalent of the cardboard tube inside the roll of paper towels—and then out into a network of collecting pipes. The total surface area of the plant's 38-inch-wide membranes would cover nearly 65 football fields.

"However," says Herd, "R.O. is the simplest part." First, the bay water must be treated before it's forced across the R.O. membranes. Pretreatment filters out suspended solids—such as scraps of seaweed, fish fry, aquatic larvae, sundry items of flotsam and jetsam. If this weren't done, the membranes would foul. "Pretreatment," says Herd, "is the challenging phase of desalination." Tampa Bay Water uses dual-stage sand filtration, in which incoming salt water flows up through two filtration cells, coarse-and fine-grained. Particulate matter larger than five microns in diameter that manages to pass through the sand filters gets caught in the cartridge filter—a collection of long, thin filters, like those used in swimming pools, which act as the R.O. membranes' safety net.

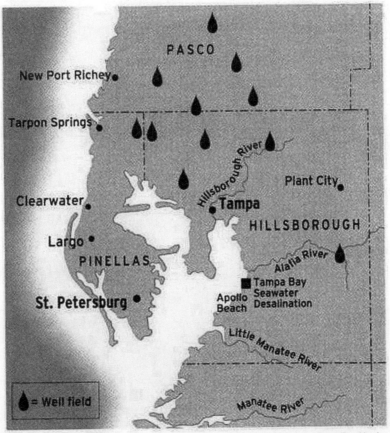

Steve Stankiewicz

143

In every performance test, both the pretreatment filters and the R.O. membranes clogged more frequently than expected, requiring additional cleaning. Increasing the strength of the cleaning solution for the membranes caused another unforeseen problem: Two million gallons of caustic, soapy cleaning fluid had to be transported to Tampa's wastewater treatment plant.

Asian green mussels turned out to be the culprit. The alien shell-fish first appeared in Tampa Bay about eight years ago, having been transported in the ballast of tankers, and has thrived. Mussels love flowing, food-rich water, so the power plant's daily 1.4 billion gallons of effluent is bivalve utopia. Larvae pass through the power plant's intake screens, survive in the heated water, then clog the pretreatment filters, fouling the R.O. membranes with microscopic hairs.

The post-treatment phase also has its complications. Along with salt, alkaloids are stripped out of the water, leaving the desalinated water acidic and corrosive to pipes. So calcium carbonate (lime) is added during post-treatment, raising the pH level before the water is piped 14 miles to storage. All this trouble and delay has resulted in lost time and money. The desal plant has declared bankruptcy three times, most recently in October 2003. The plant is online only once a month, and Tampa Bay Water says it will not go into full production until 2006.

Contemplating the sprawl that surrounds the Apollo Beach plant, I find myself paraphrasing the line from the Shoeless Joe Jackson character in the movie *Field of Dreams:* "If you build it, they will come." Herd bristles a little at the phrase. "The government agency that allows growth supports its decision with electricity, drinking water, and waste removal. The water management district doesn't have the ability to limit growth; that's the job of the planning board. Tampa Bay Water just supports the growth that's already there."

Tampa Bay Water admittedly has taken significant steps to diversify its sources of potable water, and to do so in an environmentally responsible way. As of April 2004, the water authority was pumping only 74 million gallons a day from the ailing well fields, in hopes of restoring that corner of the Floridan Aquifer. As a result of these reductions, the surrounding wetlands have begun to recover—lake levels are rising and marshland vegetation is looking fuller, more lush, a de Soto shade of green. "We didn't trade one environmental impact for another in Pasco County by shifting the burden to Tampa Bay," Herd says with justifiable pride.

Herd's optimism is refreshing. And he's right: It is not ultimately the water authority that determines the carrying capacity of a suburban landscape. Many of the 20 commercial seawater desalination projects under consideration for the Sun Belt are driven by planners who both forecast and encourage growth, often in ecologically sensitive coastal areas. Faced with lobbying by the U.S. Desalination Coalition, environmentalists will need to scrutinize each new project. For if new desal facilities mean that the wild hills become crowded with condos and the shorelines fill with sprawl, we may find ourselves echoing another line that's associated with the hero-turned-villain of the 1919 Black Sox scandal. We'll have built it, they'll have come, and like the distraught young fan, we'll be exclaiming, "Say it ain't so, Joe."

Malthus Redux: Is Doomsday upon Us, Again?

Donald G. McNeil Jr.

During the last American food-and-gas-price crisis, in the 1970s, one of my colleagues on the Berkeley student newspaper told me that he and his semi-communal housemates had taken a vote. They'd calculated they could afford meat or coffee. They chose coffee.

The decision was slightly less effete than it sounds now—the Starbucks clone wars were still some years off, so he was talking about choosing Yuban over ground chuck. But it nonetheless said something about us as spoiled Americans. Riots were relatively common in Berkeley in those days. But they were never about food. (That particular revolution was starting without us on Shattuck Avenue, where Chez Panisse had just opened.)

However, elsewhere on the globe, people were on the edge of starvation. Grain prices were soaring, rice stocks plummeting. In Ethiopia and Cambodia, people were well over the edge, and food riots helped lead to the downfall of Emperor Haile Selassie and the victory of the Khmer Rouge.

Now it's happening again. While Americans grumble about gasoline prices, food riots have seared Bangladesh, Egypt and African countries. In Haiti, they cost the prime minister his job. Rice-bowl countries like China, India and Indonesia have restricted exports and rice is shipped under armed guard.

And again, Thomas Malthus, a British economist and demographer at the turn of the 19th century, is being recalled to duty. His basic theory was that populations, which grow geometrically, will inevitably outpace food production, which grows arithmetically. Famine would result. The thought has underlain doomsday scenarios both real and imagined, from the Great Irish Famine of 1845 to the Population Bomb of 1968.

But over the last 200 years, with the Industrial Revolution, the Transportation Revolution, the Green Revolution and the Biotech Revolution, Malthus has been largely discredited. The wrenching dislocations of the last few months do not change that, most experts say. But they do show the kinds of problems that can emerge.

The whole world has never come close to outpacing its ability to produce food. Right now, there is enough grain grown on earth to feed 10 billion vegetarians, said Joel E. Cohen, professor of populations at Rockefeller University and the author of "How Many People Can the Earth Support?" But much of it is being fed to cattle, the S.U.V.'s of the protein world, which are in turn guzzled by the world's wealthy.

Theoretically, there is enough acreage already planted to keep the planet fed forever, because 10 billion humans is roughly where the United Nations predicts that the world population will plateau in 2060. But success depends on portion control; in the late 1980s, Brown University's World Hunger Program calculated that the world then could sustain 5.5 billion vegetarians, 3.7 billion South Americans or 2.8 billion North Americans, who ate more animal protein than South Americans.

Even if fertility rates rose again, many agronomists think the world could easily support 20 billion to 30 billion people.

Anyone who has ever flown across the United States can see how that's possible: there's a lot of empty land down there. The world's entire population, with 1,000 square feet of living space each, could fit into Texas. Pile people atop each other like Manhattanites, and they get even more elbow room.

Water? When it hits $150 a barrel, it will be worth building pipes from the melting polar icecaps, or desalinating the sea as the Saudis do.

The same potential is even more obvious flying around the globe. The slums of Mumbai are vast; but so are the empty arable spaces of Rajasthan. Africa, a huge continent with a mere 770 million people on it, looks practically empty from above. South of the Sahara, the land is rich; south of the Zambezi, the climate is temperate. But it is farmed mostly by people using hoes.

As the hungry riot, the ideas of a British thinker of 200 years ago get another look.

As Harriet Friedmann, an expert on food systems at the University of Toronto, pointed out, Malthus was writing in a Britain that echoed the dichotomy between today's rich countries and the third world: an elite of huge landowners practicing "scientific farming" of wool and wheat who made fat profits; many subsistence farmers barely scratching out livings; migration by those farmers to London slums, followed by emigration. The main difference is that emigration then was to colonies where farmland was waiting, while now it is to richer countries where jobs are.

Malthus's world filled up, and its farmers, defying his predictions, became infinitely more productive. Admittedly, emptying acreage so it can be planted with genetically modified winter wheat and harvested by John Deere combines can be a brutal process, but it is solidly within the Western canon. My Scottish ancestors, for example, became urbanites thanks to the desire of English scientific farmers (for which read "landlords and bribers of clan chiefs") to graze more sheep in the highlands. Four generations later, I got to mull the coffee-meat dilemma while actually living on newsroom pizza.

So it ultimately worked out for one spoiled Scottish-American. But what about the 800 million people who are chronically hungry, even in riot-free years?

Dr. Friedmann argues that there is a Malthusian unsustainability to the way big agriculture is practiced, that it degrades genetic diversity and the environment so much that it will eventually reach a tipping point and hunger will spread.

Others vigorously disagree. In their view, the world is almost endlessly bountiful. If food became as pricey as oil, we would plow Africa, fish-farm the oceans and build hydroponic skyscraper vegetable gardens. But they see the underlying problem in terms more Marxian than Malthusian: the rich grab too much of everything, including biomass.

For the moment, simply ending subsidies to American and European farmers would let poor farmers compete, which besides feeding their families would push down American food prices and American taxes.

Tyler Cowen, a George Mason University economist, notes that global agriculture markets are notoriously unfree and foolishly managed. Rich countries subsidize farmers, but poor governments fix local grain prices or ban exports just when world prices rise—for example, less than 7 percent of the world's rice crosses borders. That discourages the millions of third world farmers who grow enough for themselves and a bit extra for sale from planting that bit extra.

Americans are attracted to Malthusian doom-saying, Dr. Cowen argues, "because it's a pre-emptive way to hedge your fear. Prepare yourself for the worst, and you feel safer than when you're optimistic."

Dr. Cohen, of Rockefeller University, sees it in more sinister terms: Americans like Malthus because he takes the blame off us. Malthus says the problem is too many poor people.

Or, to put it in the terms in which the current crisis is usually explained: too many hard-working Chinese and Indians who think they should be able to eat pizza, meat and coffee and aspire to a reservation at Chez Panisse. They get blamed for raising global prices so much that poor Africans and Asians can't afford porridge and rice. The truth is, the upward pressure was there before they added to it.

America has always been charitable, so the answer has never been, "Let them eat bean sprouts." But it has been, "Let them eat subsidized American corn shipped over in American ships." That may need to change.

Global Response Required: Stopping the Spread of Nuclear Weapons

New challenges call for new approaches.

Jeffrey G. Martin and Matt Martin

When North Korea tested what is presumed to be a nuclear device on October 9, it was another blow to an already wobbly nuclear nonproliferation regime. That regime—a collection of treaties, institutions, norms, and commonly accepted practices—is the work of a nearly 40-year effort to stop the spread of nuclear weapons.

In many ways, the world's effort to keep the proliferation of nuclear weapons in check is a model of international cooperation. It addresses one of the most serious threats to humankind—the prospect of an unrestrained arms race involving uniquely powerful weapons, those with the capability of ending human life on the planet. At the same time, it holds out the promise of nuclear energy to those nations who are responsible members of the regime. By and large, the regime has served the world very well.

A Different Era

But the nonproliferation regime is also a relic of the Cold War. Its centerpiece, the Nuclear Non-Proliferation Treaty (NPT), went into force in 1970. Shaken by China's nuclear weapons development and President Kennedy's dire prediction that we could be facing a world of dozens of nuclear weapon states, the United States led the world in creating a system to control the spread of nuclear weapons. As part of the bargain, the nuclear states agreed to share nuclear energy technology so that all could benefit, and the superpowers also promised to eventually get rid of all of their nuclear weapons, while conveniently finessing the question of when they might get around to doing so.

The Cold War rivalry itself also played a role in checking the spread of nuclear weapons to other countries. Nations that allied with one of the superpowers or were de facto client states understood that their relationship with a superpower offered significant security assurances, making it easier to forgo their own nuclear weapons development. And by allowing international inspections that made sure that no nuclear materials were diverted to weapons programs, even nonnuclear weapon states could develop advanced civilian nuclear energy, medical, and research programs.

In many respects, today's world is much less orderly. The absence of a superpower rivalry limits the options for countries looking for a guarantor of their security. As well, at a time when international accountability seems to be in decline, the repercussions for a state moving to develop nuclear weapons has become much less clear. In 1962 the United States and the Soviet Union nearly started World War III when the USSR attempted to place nuclear missiles inside Cuba. Yet by 1998, two longtime adversaries, India and Pakistan, openly and repeatedly tested their own nuclear weapons, with only minor, short-term negative consequences.

Isolated and dangerous, North Korea is well on its way to becoming a full-blown nuclear weapons power. Iran, which swears it does not want nuclear weapons, nevertheless asserts its right to develop technology that would allow it to produce weapons. Without a reliable, overarching framework to hold regional and global forces in check, the "domino theory" that never came true during the Cold War may actually become a reality today.

The acquisition of nuclear weapons by states in already unstable regions might prompt their neighbors to likewise arm. The cases of North Korea and Iran have already prompted statements and speculation about South Korea, Japan, Syria, Egypt, and Saudi Arabia, and whether these states and others may react to nuclear developments in their region by adapting their own nuclear policies.

Nuclear Terrorism

Add to this the mounting concern about what radically disaffected groups who have used terrorist tactics would do if they were to get their hands on nuclear weapons.

The acquisition of nuclear weapons by states in already unstable regions might prompt their neighbors to likewise arm.

In recent years, there have been alarming indications of how this might come about. Following the breakup of the Soviet Union, thousands of nuclear weapons and thousands more tons of nuclear material were left in limbo, under shaky security. The September 11 attacks awoke the US government and public to the potential of international terrorism on US soil. And the uncovering of the A. Q. Khan proliferation network was a surprising, concrete example of how the spread of nuclear weapons and technology could be accomplished.

Energy Demands

Meanwhile, the demand for nuclear energy is growing after more than two decades of being in a stall. The rapid growth of economies and population in China, India, Brazil, and elsewhere—combined with a growing appreciation for the dangers of global warming—has created a tremendous demand for new, noncarbon-based energy sources. Nuclear energy is seen as an essential part of the mix. But an increase in the number of nations that have nuclear technology and nuclear materials raises the risk that the technology or fuel can fall into the wrong hands.

Responding to the Challenge

Independently and collectively, the global community, led by the United States, has in part responded to these challenges.

A set of programs called Cooperative Threat Reduction has secured or removed much of the "loose nukes" problem in Russia, although it will not complete its task for another decade at the current rate. The UN Security Council has passed a binding resolution making states more accountable for terrorist groups and the transit of nuclear items within their borders. And several proposals have been put forth for expanding nuclear energy without contributing to weapons proliferation.

But to date, most of our efforts have been ad hoc, bilateral, and specific. Global problems require global solutions. Will we find the vision and the political will to pursue them?

A World *Enslaved*

There are now more slaves on the planet than at any time in human history. True abolition will elude us until we admit the massive scope of the problem, attack it in all its forms, and empower slaves to help free themselves.

E. BENJAMIN SKINNER

Standing in New York City, you are five hours away from being able to negotiate the sale, in broad daylight, of a healthy boy or girl. He or she can be used for anything, though sex and domestic labor are most common. Before you go, let's be clear on what you are buying. A slave is a human being forced to work through fraud or threat of violence for no pay beyond subsistence. Agreed? Good.

Most people imagine that slavery died in the 19th century. Since 1817, more than a dozen international conventions have been signed banning the slave trade. Yet, today there are more slaves than at any time in human history.

And if you're going to buy one in five hours, you'd better get a move on. First, hail a taxi to JFK International Airport, and hop on a direct flight to Port-au-Prince, Haiti. The flight takes three hours. After landing at Toussaint L'Ouverture International Airport, you will need 50 cents for the most common form of transport in Port-au-Prince, the tap-tap, a flatbed pickup retrofitted with benches and a canopy. Three quarters of the way up Route de Delmas, the capital's main street, tap the roof and hop out. There, on a side street, you will find a group of men standing in front of Le Réseau (The Network) barbershop. As you approach, a man steps forward: "Are you looking to get a person?"

Meet Benavil Lebhom. He smiles easily. He has a trim mustache and wears a multicolored, striped golf shirt, a gold chain, and Doc Martens knockoffs. Benavil is a courtier, or broker. He holds an official real estate license and calls himself an employment agent. Two thirds of the employees he places are child slaves. The total number of Haitian children in bondage in their own country stands at 300,000. They are the *restavèks*, the "staywiths," as they are euphemistically known in Creole. Forced, unpaid, they work in captivity from before dawn until night. Benavil and thousands of other formal and informal traffickers lure these children from desperately impoverished rural parents, with promises of free schooling and a better life.

The negotiation to buy a child slave might sound a bit like this:

"How quickly do you think it would be possible to bring a child in? Somebody who could clean and cook?" you ask. "I don't have a very big place; I have a small apartment. But I'm wondering how much that would cost? And how quickly?"

"Three days," Benavil responds.

"And you could bring the child here?" you inquire. "Or are there children here already?"

"I don't have any here in Port-au-Prince right now," says Benavil, his eyes widening at the thought of a foreign client. "I would go out to the countryside."

You ask about additional expenses. "Would I have to pay for transportation?"

"*Bon,*" says Benavil. "A hundred U.S."

Smelling a rip-off, you press him, "And that's just for transportation?"

"Transportation would be about 100 Haitian," says Benavil, or around $13, "because you'd have to get out there. Plus [hotel and] food on the trip. Five hundred gourdes."

"Okay, 500 Haitian," you say.

Now you ask the big question: "And what would your fee be?" This is the moment of truth, and Benavil's eyes narrow as he determines how much he can take you for.

"A hundred. American."

"That seems like a lot," you say, with a smile so as not to kill the deal. "How much would you charge a Haitian?"

Benavil's voice rises with feigned indignation. "A hundred dollars. This is a major effort."

You hold firm. "Could you bring down your fee to 50 U.S.?"

Benavil pauses. But only for effect. He knows he's still got you for much more than a Haitian would pay. "*Oui,*" he says with a smile.

But the deal isn't done. Benavil leans in close. "This is a rather delicate question. Is this someone you want as just a worker? Or also someone who will be a 'partner'? You understand what I mean?"

You don't blink at being asked if you want the child for sex. "I mean, is it possible to have someone that could be both?"

"*Oui!*" Benavil responds enthusiastically.

If you're interested in taking your purchase back to the United States, Benavil tells you that he can "arrange" the proper papers to make it look as though you've adopted the child.

He offers you a 13-year-old girl.

"That's a little bit old," you say.

"I know of another girl who's 12. Then ones that are 10, 11," he responds.

The negotiation is finished, and you tell Benavil not to make any moves without further word from you. Here, 600 miles from the United States, and five hours from Manhattan, you have successfully arranged to buy a human being for 50 bucks.

The Cruel Truth

It would be nice if that conversation, like the description of the journey, were fictional. It is not. I recorded it on Oct. 6, 2005, as part of four years of research into slavery on five continents. In the popular consciousness, "slavery" has come to be little more than just a metaphor for undue hardship. Investment bankers routinely refer to themselves as "high-paid wage slaves." Human rights activists may call $1-an-hour sweatshop laborers slaves, regardless of the fact that they are paid and can often walk away from the job. But the reality of slavery is far different. Slavery exists today on an unprecedented scale. In Africa, tens of thousands are chattel slaves, seized in war or tucked away for generations. Across Europe, Asia, and the Americas, traffickers have forced as many as 2 million into prostitution or labor. In South Asia, which has the highest concentration of slaves on the planet, nearly 10 million languish in bondage, unable to leave their captors until they pay off "debts," legal fictions that in many cases are generations old.

Few in the developed world have a grasp of the enormity of modern-day slavery. Fewer still are doing anything to combat it. Beginning in 2001, U.S. President George W. Bush was urged by several of his key advisors to vigorously enforce the Victims of Trafficking and Violence Protection Act, a U.S. law enacted a month earlier that sought to prosecute domestic human traffickers and cajole foreign governments into doing the same. The Bush administration trumpeted the effort—at home via the Christian evangelical media and more broadly via speeches and pronouncements, including in addresses to the U.N. General Assembly in 2003 and 2004. But even the quiet and diligent work of some within the U.S. State Department, which credibly claims to have secured more than 100 antitrafficking laws and more than 10,000 trafficking convictions worldwide, has resulted in no measurable decline in the number of slaves worldwide. Between 2000 and 2006, the U.S. Justice Department increased human trafficking prosecutions from 3 to 32, and convictions from 10 to 98. By 2006, 27 states had passed anti-trafficking laws. Yet, during the same period, the United States liberated less than 2 percent of its own modern-day slaves. As many as 17,500 new slaves continue to enter bondage in the United States every year.

The West's efforts have been, from the outset, hamstrung by a warped understanding of slavery. In the United States, a hard-driving coalition of feminist and evangelical activists has forced the Bush administration to focus almost exclusively on the sex trade. The official State Department line is that voluntary prostitution does not exist, and that commercial sex is the main driver of slavery today. In Europe, though Germany and the Netherlands have decriminalized most prostitution, other nations such as Bulgaria have moved in the opposite direction, bowing to U.S. pressure and cracking down on the flesh trade. But, across the Americas, Europe, and Asia, unregulated escort services are exploding with the help of the Internet. Even when enlightened governments have offered clearheaded solutions to deal with this problem, such as granting victims temporary residence, they have had little impact.

Many feel that sex slavery is particularly revolting—and it is. I saw it firsthand. In a Bucharest brothel, for instance, I was offered a mentally handicapped, suicidal girl in exchange for a used car. But for every one woman or child enslaved in commercial sex, there are at least 15 men, women, and children enslaved in other fields, such as domestic work or agricultural labor. Recent studies have shown that locking up pimps and traffickers has had a negligible effect on the aggregate rates of bondage. And though eradicating prostitution may be a just cause, Western policies based on the idea that all prostitutes are slaves and all slaves are prostitutes belittles the suffering of all victims. It's an approach that threatens to put most governments on the wrong side of history.

Indebted for Life

Save for the fact that he is male, Gonoo Lal Kol typifies the average slave of our modern age. (At his request, I have changed his first name.) Like a vast majority of the world's slaves, Gonoo is in debt bondage in South Asia. In his case, in an Indian quarry. Like most slaves, Gonoo is illiterate and unaware of the Indian laws that ban his bondage and provide for sanctions against his master. His story, told to me in more than a dozen conversations inside his 4-foot-high stone and grass hutch, represents the other side of the "Indian Miracle."

Gonoo lives in Lohagara Dhal, a forgotten corner of Uttar Pradesh, a north Indian state that contains 8 percent of the world's poor. I met him one evening in December 2005 as he walked with two dozen other laborers in tattered and filthy clothes. Behind them was the quarry. In that pit, Gonoo, a member of the historically outcast Kol tribe, worked with his family 14 hours a day. His tools were simple, a rough-hewn hammer and an iron pike. His hands were covered in calluses, his fingertips worn away.

Gonoo's master is a tall, stout, surly contractor named Ramesh Garg. Garg is one of the wealthiest men in Shankargarh, the nearest sizable town, founded under the British Raj but now run by nearly 600 quarry contractors. He makes his money by enslaving entire families forced to work for no pay beyond alcohol, grain, and bare subsistence expenses. Their only use for Garg is to turn rock into silica sand, for colored glass, or gravel, for roads or ballast. Slavery scholar Kevin Bales estimates that a slave in the 19th-century American South had to work 20 years to recoup his or her purchase price. Gonoo and the other slaves earn a profit for Garg in two years.

Every single man, woman, and child in Lohagara Dhal is a slave. But, in theory at least, Garg neither bought nor owns them. They are working off debts, which, for many, started at less than $10. But interest accrues at over 100 percent annually here. Most of the debts span at least two generations, though they have no legal standing under modern Indian law. They are a fiction that Garg constructs through fraud and maintains through violence. The seed of Gonoo's slavery, for instance, was a loan of 62 cents. In 1958, his grandfather borrowed that amount from the owner of a farm where he worked. Three generations and three slavemasters later, Gonoo's family remains in bondage.

Bringing Freedom to Millions

Recently, many bold, underfunded groups have taken up the challenge of tearing out the roots of slavery. Some gained fame through dramatic slave rescues. Most learned that freeing slaves is impossible unless the slaves themselves choose to be free. Among the Kol of Uttar Pradesh, for instance, an organization called Pragati Gramodyog Sansthan (Progressive Institute for Village Enterprises, or PGS) has helped hundreds of families break the grip of the quarry contractors. Working methodically since 1985, PGS organizers slowly built up confidence among slaves. With PGS's help, the Kol formed microcredit unions and won leases to quarries so that they could keep the proceeds of their labor. Some bought property for the first time in their lives, a cow or a goat, and their incomes, which had been nil, multiplied quickly. PGS set up primary schools and dug wells. Villages that for generations had known nothing but slavery began to become free. PGS's success demonstrates that emancipation is merely the first step in abolition. Within the developed world, some national law enforcement agencies such as those in the Czech Republic and Sweden have finally begun to pursue the most culpable of human trafficking—slave-trading pimps and unscrupulous labor contractors. But more must be done to educate local police, even in the richest of nations. Too often, these street-level law enforcement personnel do not understand that it's just as likely for a prostitute to be a trafficking victim as it is for a nanny working without proper papers to be a slave. And, after they have been discovered by law enforcement, few rich nations provide slaves with the kind of rehabilitation, retraining, and protection needed to prevent their re-trafficking. The asylum now granted to former slaves in the United States and the Netherlands is a start. But more must be done.

The United Nations, whose founding principles call for it to fight bondage in all its forms, has done almost nothing to combat modern slavery. In January, Antonio Maria Costa, executive director of the U.N. Office on Drugs and Crime, called for the international body to provide better quantification of human trafficking. Such number crunching would be valuable in combating that one particular manifestation of slavery. But there is little to suggest the United Nations, which consistently fails to hold its own member states accountable for widespread slavery, will be an effective tool in defeating the broader phenomenon.

Any lasting solutions to human trafficking must involve prevention programs in at-risk source countries. Absent an effective international body like the United Nations, such an effort will require pressure from the United States. So far, the United States has been willing to criticize some nations' records, but it has resisted doing so where it matters most, particularly in India. India abolished debt bondage in 1976, but with poor enforcement of the law locally, millions remain in bondage. In 2006 and 2007, the U.S. State Department's Office to Monitor and Combat Trafficking in Persons pressed U.S. Secretary of State Condoleezza Rice to repudiate India's intransigence personally. And, in each instance, she did not.

The psychological, social, and economic bonds of slavery run deep, and for governments to be truly effective in eradicating slavery, they must partner with groups that can offer slaves a way to pull themselves up from bondage. One way to do that is to replicate the work of grassroots organizations such as Varanasi, India-based MSEMVS (Society for Human Development and Women's Empowerment). In 1996, the Indian group launched free transitional schools, where children who had been enslaved learned skills and acquired enough literacy to move on to formal schooling. The group also targeted mothers, providing them with training and start-up materials for microenterprises. In Thailand, a nation infamous for sex slavery, a similar group, the Labour Rights Promotion Network, works to keep desperately poor Burmese immigrants from the clutches of traffickers by, among other things, setting up schools and health programs. Even in the remote highlands of southern Haiti, activists with Limyè Lavi ("Light of Life") reach otherwise wholly isolated rural communities to warn them of the dangers of traffickers such as Benavil Lebhom and to help them organize informal schools to keep children near home. In recent years, the United States has shown an increasing willingness to help fund these kinds of organizations, one encouraging sign that the message may be getting through.

For four years, I saw dozens of people enslaved, several of whom traffickers like Benavil actually offered to sell to me. I did not pay for a human life anywhere. And, with one exception, I always withheld action to save any one person, in the hope that my research would later help to save many more. At times, that still feels like an excuse for cowardice. But the hard work of real emancipation can't be the burden of a select few. For thousands of slaves, grassroots groups like PGS and MSEMVS can help bring freedom. But, until governments define slavery in appropriately concise terms, prosecute the crime aggressively in all its forms, and encourage groups that empower slaves to free themselves, millions more will remain in bondage. And our collective promise of abolition will continue to mean nothing at all.

E. Benjamin Skinner is the author of *A Crime So Monstrous: Face-to-Face with Modern-Day Slavery* (New York: Free Press, 2008).

Test-Your-Knowledge Form

We encourage you to photocopy and use this page as a tool to assess how the articles in *Annual Editions* expand on the information in your textbook. By reflecting on the articles you will gain enhanced text information. You can also access this useful form on a product's book support website at *http://www.mhcls.com*.

NAME: DATE:

_____

TITLE AND NUMBER OF ARTICLE:

_____

BRIEFLY STATE THE MAIN IDEA OF THIS ARTICLE:

_____

LIST THREE IMPORTANT FACTS THAT THE AUTHOR USES TO SUPPORT THE MAIN IDEA:

_____

WHAT INFORMATION OR IDEAS DISCUSSED IN THIS ARTICLE ARE ALSO DISCUSSED IN YOUR TEXTBOOK OR OTHER READINGS THAT YOU HAVE DONE? LIST THE TEXTBOOK CHAPTERS AND PAGE NUMBERS:

_____

LIST ANY EXAMPLES OF BIAS OR FAULTY REASONING THAT YOU FOUND IN THE ARTICLE:

_____

LIST ANY NEW TERMS/CONCEPTS THAT WERE DISCUSSED IN THE ARTICLE, AND WRITE A SHORT DEFINITION:

We Want Your Advice

ANNUAL EDITIONS revisions depend on two major opinion sources: one is our Advisory Board, listed in the front of this volume, which works with us in scanning the thousands of articles published in the public press each year; the other is you—the person actually using the book. Please help us and the users of the next edition by completing the prepaid article rating form on this page and returning it to us. Thank you for your help!

ANNUAL EDITIONS: Geography 23/e

ARTICLE RATING FORM

Here is an opportunity for you to have direct input into the next revision of this volume.
We would like you to rate each of the articles listed below, using the following scale:

1. **Excellent: should definitely be retained**
2. **Above average: should probably be retained**
3. **Below average: should probably be deleted**
4. **Poor: should definitely be deleted**

Your ratings will play a vital part in the next revision.
Please mail this prepaid form to us as soon as possible.
Thanks for your help!

RATING	ARTICLE
	1. The Big Questions in Geography
	2. Rediscovering the Importance of Geography
	3. The Four Traditions of Geography
	4. The Changing Landscape of Fear
	5. The Geography of Ecosystem Services
	6. The Agricultural Impact of Global Climate Change: How Can Developing-Country Farmers Cope?
	7. When Diversity Vanishes
	8. Classic Map Revisited: The Growth of Megalopolis
	9. A Great Wall of Waste
	10. In Niger, Trees and Crops Turn Back the Desert: A Poor African Nation Uses a Simple Mix to Grow Greener
	11. Whither the World's Last Forest?
	12. Why It's Time for a "Green New Deal"
	13. Study Finds Humans' Effect on Oceans Comprehensive
	14. Polar Distress
	15. What China Can Learn from Japan on Cleaning up the Environment
	16. The Rise of India
	17. Hints of a Comeback for Nation's First Superhighway
	18. The Short End of the Longer Life
	19. Never Too Late to Scramble: China in Africa
	20. Where Business Meets Geopolitics
	21. Half-Way from Rags to Riches

RATING	ARTICLE
	22. Malaria: It's Not Neglected Any More (But It's Not Gone, Either)
	23. Tsunamis: How Safe Is the United States?
	24. The World Is Spiky
	25. Hurricane Hot Spots: Most Vulnerable Cities
	26. Sea Change: The Transformation of the Arctic
	27. Shaping the World to Illustrate Inequalities in Health
	28. Deaths Outnumber Births in Third of Counties
	29. How Much Is Your Vote Worth?
	30. *Fortune* Teller
	31. Teaching Note: The U.S. Ethanol Industry with Comments on the Great Plains
	32. Clogged Arteries
	33. Manifest Destinations
	34. AIDS Infects Education Systems in Africa
	35. Wonderful World? The Way We Live Now
	36. What Lies Beneath
	37. Cloud, or Silver Linings?
	38. Troubled Waters
	39. Turning Oceans into Tap Water
	40. Malthus Redux: Is Doomsday upon Us, Again?
	41. Global Response Required: Stopping the Spread of Nuclear Weapons
	42. A World *Enslaved*

BUSINESS REPLY MAIL
FIRST CLASS MAIL PERMIT NO. 551 DUBUQUE IA

POSTAGE WILL BE PAID BY ADDRESSEE

McGraw-Hill Contemporary Learning Series
501 BELL STREET
DUBUQUE, IA 52001

ABOUT YOU

Name   Date

_____

Are you a teacher? ❏ A student? ❏
Your school's name

_____

Department

_____

Address   City   State   Zip

_____

School telephone #

_____

YOUR COMMENTS ARE IMPORTANT TO US!

Please fill in the following information:
For which course did you use this book?

_____

Did you use a text with this ANNUAL EDITION? ❏ yes ❏ no
What was the title of the text?

_____

What are your general reactions to the Annual Editions concept?

_____

Have you read any pertinent articles recently that you think should be included in the next edition? Explain.

_____

Are there any articles that you feel should be replaced in the next edition? Why?

_____

Are there any World Wide Websites that you feel should be included in the next edition? Please annotate.

_____

May we contact you for editorial input? ❏ yes ❏ no
May we quote your comments? ❏ yes ❏ no